WHY WHALES SING

WHY
WHALES
SING

EDUARDO MERCADO III

JOHNS HOPKINS UNIVERSITY PRESS

Baltimore

© 2025 Johns Hopkins University Press
All rights reserved. Published 2025
Printed in the United States of America on acid-free paper
9 8 7 6 5 4 3 2 1

Johns Hopkins University Press
2715 North Charles Street
Baltimore, Maryland 21218
www.press.jhu.edu

Library of Congress Cataloging-in-Publication Data is available.

A catalog record for this book is available from the British Library.
ISBN 978-1-4214-5289-0 (hardcover)
ISBN 978-1-4214-5290-6 (ebook)

Special discounts are available for bulk purchases of this book. For more information, please contact Special Sales at specialsales@jh.edu.

EU GPSR Authorized Representative
LOGOS EUROPE, 9 rue Nicolas Poussin,
17000, La Rochelle, France
E-mail: Contact@logoseurope.eu

CONTENTS

PROLOGUE

The world of whales is an idea rather than a place. People (including scientists) can describe how a whale's world appears from only a human perspective. For most of human history, that imagined watery world was a fantasy extrapolated from personal experiences and borrowed beliefs about the way worlds should be. Only a few luminaries like Aristotle actively explored what the worlds of cetaceans—whales, dolphins, and porpoises—were really like. Aristotle flung dolphins out of the fish family and into the mammalian clan, noting that "this creature has a voice" (*Historia Animalium*, book IV, pt. 9).

Aristotle set the stage for subsequent scientific studies of animal sounds by pointing out that "all animals without exception exercise their power of singing or chattering chiefly in connexion with the intercourse of the sexes" (*Historia Animalium*, book I, pt. 1). People and other animals chatter for purposes beyond intercourse, of course. But like many before and after him, Aristotle did not view humans as animals. He also had only a sketchy understanding of how cetaceans sense the world. Aristotle argued correctly that dolphins can hear, see, and feel but also claimed they have an exquisite sense of smell (most dolphins can't smell squat). Aristotle certainly didn't know that cetaceans can hear sounds so high-pitched that humans cannot perceive them at all. Nor did he suspect that cetaceans can use such sounds to generate echoes that enable them to recognize objects across long distances in complete darkness.

Historically, philosophers have adopted a rather dim view on the mental experiences of animals. Descartes famously argued that everything nonhuman was mindless, implying that animals' worlds are unlike anything because they, like rocks, lack awareness. As psychologists and neuroscientists began actively experimenting with animals in the twentieth century, however, evidence accumulated that the mental states of humans are not as unique as one might think based on their impressive linguistic and technological accomplishments.

Scientists have made great strides in understanding brains and behavior. Still, human biases and preconceived notions continue to keep people from fully appreciating how other animals may perceive themselves and their surroundings. Such biases determine what scientists study and how they interpret their findings. On the one hand, common sense could tell you that if two animals both have pairs of eyes connected to similar brain circuits, then there's a decent chance they have a comparable capacity to see. It was just that sort of reasoning that led scientists to overlook the fact that some animals can see wavelengths of light that humans can't (specifically, infrared and ultraviolet), as well as colors that humans have no ability to distinguish or imagine. Scientists and philosophers initially missed this because they assumed human visual abilities were as good as it gets.

Around the same time that technological advances made it possible for scientists to test animals' sensitivities to "imperceptible" lights, existential philosophers such as Heidegger promoted the idea that an individual's thoughts and experiences are shaped by the environments within which they exist. According to Heidegger, one's activities are a key component of "being-in-the-world" because of the world's dynamic and evolving nature. When organisms play an active role in creating their environments—for example, by inventing cities and cell phones—then the dynamic interactions among perceptual experiences, actions, and environmental conditions can come to define an individual's world.

Heidegger was talking about humans, but his ideas also apply to animals that echolocate in darkness. Their vocal actions and bodily movements effectively create the environments within which they navigate, especially when they're moving through unfamiliar terrain. For a cetacean swimming near the surface at night, there is no "ocean floor" aside from the one it perceives by listening to returning echoes. Likewise, they also perceive no predators, prey, or obstacles, and in many cases no peers, aside from the echoes that bounce from others' bodies.

Certain cetaceans and bats are the best-known echolocators. A few other animals also share the ability to find their way using echoes. Swiftlets, birds famous for building the nests that humans eat in bird's nest soup, navigate through dark caves using echoes from the clicks they produce. Some blind humans also move through their worlds (and even get around on bikes!) by making clicks and learning to interpret the resulting echoes. Human echolocators describe sensing the shape, size, texture, identity, distance, and movements of objects around them, as well as "sound shadows" that happen when a nearby object blocks echoes.

This book focuses on "singing" whales. Specifically, it closely considers what whales are doing when they produce complex vocal patterns—patterns that to human ears resemble melodic songs. Over the past 30 years, I've explored the possibility that whales "sing" to echolocate over long distances, a possibility that most experts in the field consider preposterous. Here, I lay out the evidence that has compelled me to challenge current beliefs about why whales sing. Every chapter is an exploration of how cetaceans construct (or may construct) perceptual experiences using echoes.

Despite that fact, none of the chapters summarizes what scientists currently know about the nature or mechanisms of echolocation. I provide a brief overview of the basics of bat and dolphin echolocation in an appendix for those who are curious. But I purposefully avoided reviewing findings on bat and dolphin echolocation in the main text for one simple reason: While ignorance about ultrasound led scientists to discount early evidence of echolocation in bats, subsequent studies of ultrasonic echolocation in bats and dolphins have led many modern scientists to ignore the echoic potential of any vocalizations unlike those used by bats and dolphins.

The fact that bats and dolphins echolocate using ultrasonic vocalizations in no way precludes the possibility that they or other species might also be able to do so using audible sounds—sounds that may differ considerably in form from those used to generate ultrasonic echoes. These vocalizations may, on first, second, or third inspection, seem quite songlike.

The following chapters are an attempt to clarify the relationships between whales, sounds, and their worlds. The scientific story builds on methods and theories from several disciplines, including psychology, acoustics, neuroscience, engineering, and ecology. I've tried to avoid presenting a deluge of factoids, equations, and arcane plots, as well as the customary oceanic slide show of whales and whale researchers. In each chapter I've also provided boxes with more in-depth discussions of whale song acoustics and cross-species comparisons. A list of pertinent publications at the end of the book is provided for those keen to go all in on understanding the science of whale song. I've even included an interview with a dolphin echolocation expert (in chapter 5) to provide a more mainstream scientific perspective on whale song.

One last note on terminology. The term "whales" appears many times throughout this book. When I write "whales," I'm referring to a particular subset of animals, specifically the Mysticeti or whalebone whales. These include humpback whales, the cetaceans best known for their dramatic leaps and enigmatic songs. Technically, the term "whales" refers to all those marine mammals with both flippers and flukes. That includes dolphins and porpoises. When I refer to "singing whales," what I mean are those Mysticeti known to sing (revealed in chapter 3). Only singing humpback whales (*Megaptera novaeangliae*) have been studied in any detail, however. So in most cases "singing whales" means singing humpback whales. When it means something else—like singing bowhead whales—I'll let you know.

This is not a book that summarizes scientists' greatest hits. It's a book about the scientific process—the good, the bad, and the ugly. Here you'll find new hypotheses, old evidence, and surprising suppositions. Explanations for what whales are doing and why will increase in complexity as you progress through chapters, so be prepared. Science is seldom easy. I'll do my best to help you connect the dots so you can decide for yourself what makes sense and what's fanciful. I'm assuming that if you're reading this book, it's because you're curious about what animals can do. Me too! I wrote this book for a party of two—me and you. There's so much more to discover, so many facets that remain unexplored. One chance encounter, one serendipitous discovery or insight, could change the way you see the world. So keep your eyes, ears, and mind open, and let me tell you a few things I've learned about how whales and science work.

WHY WHALES SING

CHAPTER 1

Why Whales Sing

WARBLES THAT RISE AND SLOWLY FALL, the staccato hum of a motor running, pairs of wavering tones followed by faint pulses, elaborate harmonic arpeggios, a great variety of squeaks and chirps, a soaring warbling note, a deepening series of roars or bellows: This is how scientists first described the sounds they heard from submerged whales. It's no wonder that to many human ears, these haunting vocalizations are reminiscent of songs.

The earliest recordings of vocalizing whales were not made by scientists, however, but by military surveillance specialists. We don't really know when these first recordings were made or when it became clear that the sounds were coming from whales, because the US Navy tends not to divulge such details. We do know when scientists first became aware that the sounds are surprisingly structured—a fact not discovered by scientists, either, but by an English major and his math teacher wife after staring at long printouts of visualized audio recordings laid out on their living-room floor. The couple inherited this task from a biologist named Roger Payne, who had been given recordings of humpback whales collected in Bermuda. Upon careful inspection of the printouts, the couple concluded, "My God. It repeats." By printing out the patterns, they discovered that humpbacks repeat complex sequences of sounds in a fixed order, like an album on a loop. It was Roger Payne, in collaboration with his wife, zoologist Katy Payne,[1] who dubbed the repetitions *songs*. Their decision to describe the repeating patterns as songs strongly shaped how scientists would study whale sounds in the coming decades.

What is it about singing animals that people find appealing? Part of the answer is that certain sound patterns trigger the release of neurochemicals in humans that lead to feelings of pleasure. Romantic, right? Additionally, rhythmic patterns can cause neural circuits to dance in synchrony, which can lead to arousal and feelings of happiness. In the case of humpback whale songs, it was the idea of whales singing like some voluptuous sirens that captured peoples' imaginations. At the same time, public awareness of the mass slaughter of these talented mermammals was growing. Singing whales were transformed from a scientific surprise into a meme—a story around which environmental activists and animal lovers could unite.

The year I was born, 1970, was also when the first scientific descriptions of whale songs were published. Coincidentally, the first songs described were recorded off Puerto Rico, the birthplace of my father and grandparents. Two of those songs were recorded during my birth week! But I wouldn't hear a whale song until decades later, as a young scientist in the 1990s. Upon my first listening, I found whale songs to be neither impressive nor appealing. In contrast, this is what Katy Payne reported about the first time she and Roger heard singing humpbacks: "Oh, my God, tears flowed from our cheeks. We were just completely transfixed and amazed because the sounds are so beautiful, so powerful."[2]

---

1. Katy Payne is now better known for her groundbreaking studies of elephant vocalizations and for founding Cornell University's Elephant Listening Project.

2. Bill McQuay and Christopher Joyce, "It took a musician's ear to decode the complex song in whale calls," *CapRadio*, August 6, 2015, https://www.capradio.org/news/npr/story ?storyid=427851306.

To me, singing humpback whales sounded more like cows that had eaten some bad grass than like beautiful balladeers. These different impressions may be due to generational or disciplinary differences. Researchers listening to whales in the 1970s were boat-based biologists sailing through crystal clear waters to explore their hidden depths. My initial exposure to whale song was as a computer-anchored psychologist forced to sort and catalog archived tape recordings. When biologists study singing whales, they are listening for evolutionary origins, social scenarios, and ornamental extravagances. When psychologists[3] study singers, they are more likely to listen for linguistic features, signs of memory and attention, or symptoms of clinical depression. Where biologists see beauty, psychologists see routines.

Given my lackluster reaction to the hypnotic harmonies of humpback whale songs, you may be wondering why this book exists. Why would someone write an entire book about gargling mammals? My views on humpback whale song changed not because I began to hear them differently but because of a few pages I happened to read in a book written by a scientist who would later become one of my most vocal critics. Those few pages provoked me to think more deeply about what it's like to be a singing whale and why whales sing. I give the author full credit for that, although I'm not sure he'd appreciate being recognized as the trigger for the shift in thinking he sparked.

Do Whales Sing?

Whales have been producing complex sound sequences for longer than humans have existed. The fact that humans recognize patterns in these sequences is what qualifies them as songs. Scientists describe repetitive patterns produced by birds, frogs, and

---

3. Do psychologists study singing whales? So far, there have been five of us.

insects as songs, so why not whales? This kind of criterion-based classification seems logical at first glance. But how consistently do scientists classify repetitive sound patterns as songs? Dogs often bark in recognizable patterns, but most are not described as singing. Prairie dogs[4] also produce series of calls in predictable patterns, but they, too, are not deemed singers. People taking a bathroom break produce recognizably patterned sequences of sounds, typically ending with a flushing-washing-drying refrain, but no scientists describe these toilet tunesmiths as singers. Clearly, more is required before scientists classify a set of sounds as a song. In almost all cases, the deciding factor is the context. Specifically, if animals produce repetitive patterns in situations where listeners may be interested in the no-pants dance, then scientists are much more likely to describe those patterns as songs. Humpback whales were initially recorded in exactly this scenario.

The first clue that humpback whales vocalizing in the tropics were males attempting to mate came from their size. Male humpbacks tend to be smaller than females, and some early observations suggested singers were smaller. Researchers' initial interpretations may have been biased, however, by what they knew of other species. In the first paper detailing the properties of whale songs, the authors write, "In other species of animals, it is usually the males that 'sing' during the mating season. If we are to ascribe the usual functions of such calling, it would be unusual if it were the female." Later inspections of genitals and genetics confirmed that male whales sing, and now most whale researchers confidently claim that *only* males sing. For the moment, let's be a bit more conservative and say that lots of male whales sing. And when they do sing, they do it a lot. Given those facts, most whale researchers (including me) believe that singing is a way for males to better

---

4. Not a dog.

their chances with the ladies. Where scientists part ways is in their beliefs about how these songs help singers get busy.

Currently, the most popular explanation for why whales sing is that songs are a way for a male to communicate to choosy females why they should mate with him and a way to let other males know that he's on the prowl.[5] In other words, it's a mating display, akin to the showy tail of a male peacock. Many whale researchers find this portrayal of singing whales as aquatic peacocks with exotic acoustic plumage appealing.

But it's not the only proposed explanation for why whales sing. In *Among Whales*, Roger Payne (1995) describes no fewer than eight possibilities, including that songs could be whale histories, mantras, maps, or a form of *change ringing*, which is a kind of bell-ringing orgy.[6] Each of these hypotheses for why whales sing accounts for subsets of the available evidence. Some of the explanations are not testable, meaning that there is no known way to conduct experiments or collect observations that would reveal whether the explanation is right or wrong. For those hypotheses that are testable, it's not always clear when an explanation has passed or failed a particular test. Making matters trickier, the people who judge the outcomes of different tests are usually the same people that proposed the explanations being tested and so are far from impartial. Most scientists aspire to be objective, but the sad reality is that (subjectively) none really are.

I personally find the peacock explanation for why whales sing to be a decent first guess but one that ignores much of what is known about songs and singers and whales and oceans and the results of pretty much every experiment ever designed to test this hypothesis. Few scientists share my view, so it's worth reviewing

---

5. From this perspective, singing whales are like contestants on *American Idol*, doing their best to beat out the competition and thereby win the prize—in this case, the prize being intromission. So maybe more like *The Bachelor*.

6. Payne personally prefers the popular peacock proposal.

briefly why the peacock story remains so popular: (1) Male humpback whales sing, (2) they sing the most during the time of year when most breeding takes place, (3) sometimes they sing when accompanying a female whale, and (4) songs are complex. These facts led researchers to propose that whale songs are mating displays in the 1970s, and these facts are the ones most often noted as supporting the peacock parallel 50 years later.

Scientists have learned a lot about humpback behavior and whale songs since the 1970s. In my opinion, what we know now raises serious doubts about the birdiness of humpback whales. One red flag is the amount of interest female whales show in singing males, which is essentially none. Females are rarely with singers, nor do they often approach singers. They are more likely to actively avoid them. Unlike peacocks, the luckiest of whom win a harem of females within a defendable territory, singing whales neither collect females nor defend a territory and are most often alone when singing.[7]

Another head scratcher is that singing males seem okay with other males stopping by to visit during performances. If whale songs are directed at females, it's odd that males are the ones attracted by them. In addition, the visits tend to be mellow—a male will swim over to be with a singer for a few minutes, then head off to do other things. Sometimes the two even swim off together! It's almost like a fisherman stopping by to see if another fisherman has caught anything. That lack of aggression between a singer and encroaching males is decidedly un-peacocky.

---

7. They do not form family units either and are rarely seen with the same female for more than a couple of days. In short, male whales are players, not providers.

Opposite page: *Impressing the ladies.*
A male humpback whale will sometimes sing while accompanying a mother and calf. Some researchers speculate that the singer is attempting to convince the mother to choose him as a mate, much like peacocks display their train feathers to seduce peahens.

HOW BIRDY ARE WHALES?

Biologists have characterized whales as "songbirds of the sea" since the earliest scientific reports of their vocal actions. But how birdy are whales really? The first published comparison between a songbird and a humpback whale was made by a grad student named Peter Tyack in 1981. Tyack, who has since become a top expert on whale and dolphin vocal behavior, compared humpback songs to those of canaries, noting that since female canaries favor singers with larger song reper- toires, maybe female humpbacks favor singers that produce more complex songs. This idea continues to be popular, with researchers Katie Kowarski and colleagues most recently claiming that "many aspects of the humpback whale song ontogeny align with that of songbirds, with the canary . . . being a notable example."

Despite their popularity, comparisons between singing humpbacks and singing canaries are iffy at best. Canaries sing an enormous variety of short-duration songs (usually less than 20 seconds long) made up of a set of around 40 sounds in different combinations. They may sing hundreds of different songs every year. Male canaries form pair bonds with their mates, often before the breeding season even starts—canaries are some of the most monogamous birds on the planet.

In contrast, humpback whales typically have a repertoire of one song per year. That song can last 20 minutes or more, and the sounds within it vary in lots of different ways. Male hump- back whales do not bond with mates and are not monogamous. In short, canary songs differ from whale songs in almost every way imaginable, and the circumstances in which the two species produce and use songs are also quite different. Singing canaries are like singing whales only in that they are both vertebrates that make a variety of sounds in reproductive contexts and change the songs they produce over time.

It's true that if you speed humpback whale songs up quite a bit, they sound like birds singing. Of course, if you speed their songs up even more, they sound like squeaky wheels on a tricycle. So this might not be the best strategy for finding meaningful similarities.

* *Ontogeny* refers to the process of individual development.

Other discoveries about the actions of singing and nonsinging whales, described in later chapters, make the standard explanation for why whales sing seem curiouser and curiouser. These behavioral oddities become less odd, however, if one gives up on the idea that whales are singing. By that I don't mean denying that whales are rhythmically vocalizing but abandoning the assumption that their vocalizations are ballads, serenades, or bro chants.

If the answer to "Do whales sing?" is no, then what else might they be doing? *Cetaceans*—the group that includes whales, dolphins, and porpoises, pronounced "suh-tay-shuns"—are known to produce sounds for reasons other than just expressing emotions and communicating information. They also explore their surroundings with sound.[8] Using an impressive form of auditory perception called echolocation (or biosonar), cetaceans listen to how the sounds they make reflect off objects in their environment. For example, dolphins echolocate when searching for food or while navigating around obstacles. To date, the simplest alternative hypothesis that can explain most of the behavioral and acoustic evidence collected in the last 50 years is that humpback whale "song" is a form of long-range echolocation that enables singers to detect large targets from long distances: the sonar hypothesis.

8. Their actions can also create functionless sounds, like when a whale blows after surfacing or slaps the water with its body.

The possibility that some of the sounds produced by singing humpback whales might be used for echolocation was first proposed in the 1970s by Howard and Lois Winn, another science power couple whose views on whales were sparked by hearing humpbacks singing in the tropics. The Winns were the first to note that some sounds within songs generated stronger echoes from submerged banks and peaks. They were also the first to suggest that a singer might be using echoes from its songs to determine its depth as it surfaced to breathe. They did not think this was why whales sang, however. Like other whale biologists, they viewed whale songs as sexual displays that functioned like birdsong. Their observations about song-generated echoes were largely ignored and soon forgotten.

Part of the reason why scientists discounted the sonar hypothesis early on is that the best-known underwater echolocators, dolphins, use sounds that are nothing like whale song. The sounds dolphins use to echolocate are short, high-pitched clicks—sometimes higher than humans can hear—produced in rapid succession. Whale songs did not sound like dolphin clicks or any other known kinds of echolocation, so most scientists studying cetaceans didn't think of songs as potential sonar signals.[9]

Dolphins are not the only echolocators in the animal kingdom, however. Many species of bats use sounds other than clicks to echolocate, so there is no reason to think that only clicks can generate useful echoes. The form biosonar takes depends on how and where and for what it's being used (yearning to know more about bat and

---

9. You can think of this as the flipped-duck test. The classic duck test is "If it looks like a duck, swims like a duck, and quacks like a duck, then it's probably a duck." The flipped version—for singing humpback whales—is "If it doesn't look like a dolphin echolocating, doesn't use ultrasonic clicks like a dolphin, and doesn't rapidly repeat sounds with timing like a dolphin, then it's probably not echolocating." What the flipped-duck (dolphin?) test overlooks, however, is that there is more than one kind of sonar, and the form that echolocation takes depends on how and where it's being used.

dolphin echolocation? Then head to the appendix). Singing humpback whales are not using sound to detect and catch fish, as dolphins do, or to track and capture flying insects, like bats. They are actively monitoring their surroundings over vast distances, particularly the movements of other whales.

Assume for the sake of argument that singing whales are echolocating. If you consider what a whale might hope to gain from generating slow streams of echoes, then you're likely to conclude that the echoes probably provide the whale with some information about objects in their environment. The ocean surface and bottom, as well as sunken banks, can generate strong echoes. This kind of information could help whales know their location.[10] The next most massive reflectors of echoes in the ocean are animals, with whales being the biggest.

Male humpback whales show many signs of searching to join other whales (both males and females) during the breeding season. For sure, males and females must converge to mate, and they don't normally hang out together. Some humpback whales are clearly attempting to make social contacts when they are in tropical breeding grounds. Sound may be what makes getting together possible. Sound can help cetaceans spatially converge in two ways: (1) When cetaceans make sounds, others can potentially find them, Marco-Polo style, and (2) when sounds bounce off cetaceans, listeners can potentially use those echoes to home in on the location of the "silent" animal.

You may have gotten the misimpression from the discussion so far that scientists have yet to reach a consensus about whether humpback whales use sonar. Not so. The near-unanimous consensus is that large whales do not echolocate using any sounds, much less songs. Google "Do whales echolocate?," and you will find this

---

10. Some researchers think maybe whales use these kinds of echoes to find their way around while migrating, but nonsingers also may use other clues to navigate.

Song as sexual signal

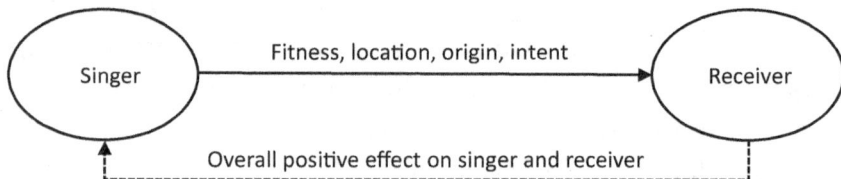

Singer — Fitness, location, origin, intent → Receiver

Overall positive effect on singer and receiver

Song as sonar

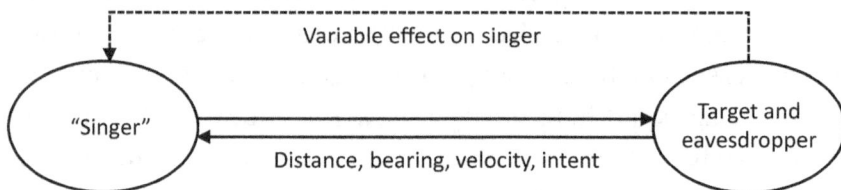

Variable effect on singer

"Singer" ⇄ Target and eavesdropper

Distance, bearing, velocity, intent

When a whale produces a song, it may provide other whales with information about the singer, such as the singer's quality, goals, and movements. When sounds within songs bounce off other whales as echoes, they can also provide the singer with relevant information about other whales.

fact noted on many sites, often in comparisons between dolphins and larger whales. If you're wondering why everyone thinks larger whales aren't echolocating,[11] you will find answers like "They do not need to echolocate because their food is little." The more open-minded sites say that humpback whales are not known to echolocate but are known to use songs for communicating.

It's true that any sound a whale produces communicates information to listeners; this also holds true for the sounds made by echolocating bats and dolphins. It's also true that any sound a whale produces generates echoes that could provide information to listeners, including the vocalizer. This is why echolocation is

---

11. I'm wondering too.

sometimes referred to as autocommunication—the producer of the signal is also the intended receiver, kind of like when people give themselves a pep talk. Surprisingly, there is no more evidence that humpback whales produce sounds to purposively communicate information to other whales than there is that they produce sounds to communicate with themselves.

It seems obvious that if one whale makes a sound and another whale reacts to it, then the first whale made the sound for the purpose of exchanging information with the second whale. The questionable nature of this assumption becomes obvious when you consider that scientists use whales' vocalizations to track them. Does this imply that whales vocalize to reveal their movements to scientists? Some bats will silently follow an echolocating bat and then swoop in at the last minute to steal an insect the echolocator was about to capture. Would you interpret this as evidence that the echolocating bat was inviting other bats to eat his dinner?

Undoubtedly, if singing humpback whales are using their complex sound sequences as a means of echoically perceiving the actions of other whales located miles away, then it's one of the most sophisticated forms of echolocation in the animal kingdom. But this capacity does not require any fantastical physics or neural processing beyond what marine mammals are already known to possess. Echolocating humpback whales may, however, have adapted their acoustic abilities in ways that differ somewhat from the specializations seen in dolphins or bats, so understanding how humpback whales use sound may require thinking about the underwater lives of whales from new perspectives.

Humpback whales have adapted to their acoustic environments in ways that dolphins have not because they have different perceptual needs. Rethinking past assumptions about how whales use sound can not only provide alternative interpretations for old observations but can also explain some mysterious (and largely ignored) properties of the sounds themselves. To top it off, the sonar hypothesis leads to novel predictions about how whale brains change over time, which could have implications for understanding how your own brain[12] works.

Plastic Whales

My desire to understand how cetacean brains and minds work emerged from a kind of existential crisis I experienced while working as a neophyte computer engineer at IBM in the early 1990s. As an undergraduate I majored in computer engineering mainly to raise my job prospects. In that respect, getting hired at IBM was ideal. I was interested in developing computers more flexible than IBM's standard models, computers that functioned more like brains. I started investigating what scientists knew about how different brains work, which eventually led me to the enigmatic brains of cetaceans. My general impression was that no one really knew what a dolphin's brain could or couldn't do, which sparked my curiosity. Cetacean brains differ from human brains in many ways. I wanted to know which differences determined what those brains could do.

I found it hard to reconcile my training and career opportunities with my emerging scientific interests in cetacean brains. The possibility that I was simply going off the deep end seemed uncomfortably likely. Ultimately, I used the empirical evidence of research papers scattered around my apartment and time logged in the zoology and neuroscience sections of university libraries to conclude that it was time for me to meet some new mammals.

Experimentally savvy dolphins and scientists who knew about dolphin cognition and communication were in rare abundance at the University of Hawai'i. I applied to grad school there, and only there, in the hopes of joining the Kewalo Basin Marine Mammal

_____

12. And mind?

Laboratory. The lab at Kewalo Basin was then famous for its language-trained dolphins. These animals could interpret a gestural sign language well enough to follow instructions to take a ball on their left and put it under a stream of water on their right (when there were balls and water on both sides of them). My interest in the Kewalo Basin dolphins had nothing to do with their linguistic capacities, however. I was more interested in their ability to learn and perform tasks that require complex cognitive skills, skills like imitating actions and sounds and recognizing objects via vision and echolocation. I wanted to know what made it possible for dolphins to learn concepts that most animals other than apes seemed unable to master. I also wanted to discover whether something about dolphins' big brains gave them exceptional conceptual powers. These are the kinds of questions that drive an engineer to become a psychology professor.

My entry into whale song science was the result of an accidental alignment. The scientific director at Kewalo Basin, Louis (Lou) Herman, originally recruited me to assist with ongoing studies of dolphin cognition. At the lab I was taught to train dolphins and eventually to train people to train dolphins. I was not training dolphins to perform fancy flips, however, but to perform more mentally challenging tasks like recognizing analogies and recalling recent episodes. At the same time, Lou was also leading field studies of humpback whale behavior in Hawai'i, including studies of singers.

Lou had recently obtained money to support a collaboration with IBM signal-processing engineers to develop algorithms for analyzing humpback whale songs. Unfortunately, no one in the lab was working on song analyses at the time and, consequently, no one was available to interact directly with the engineers. Since I had been working as an engineer at IBM before arriving in Hawai'i, Lou turned to me as a possible liaison despite me knowing essentially nothing about whales, whale song, or the humpback whale research project. What I did know was that living in Hawaiʻi was expensive and that the grant would fund my position, so I immediately said yes to the gig. The predictable result was a case of the blind leading the blind. I would dig through archived recordings, find some that sounded clear to me, and then deliver these to the IBM engineers so they could work their magic. The results ultimately proved not so magical because the engineers were attempting to perform speech analyses on the whale sounds. In order to have any sort of coherent discussion with the engineers about what was or wasn't working, I had to give myself a crash course in whale song sounds and bioacoustic analyses. This mainly involved reading stacks of scientific papers, listening to hours of recordings, and constantly pestering Adam Frankel, one of the key researchers collecting recordings of whale song in Hawaiʻi at that time.

The basic goal of the IBM collaboration was to develop a kind of dictionary of the sounds singing humpback whales make that researchers could use to rapidly sort and transcribe the complex patterns within songs. Such a dictionary would make it possible to objectively compare songs produced by different singers in different times and places. About a year into this project, it became clear to me that no such dictionary was going to work out. Each time we added recordings from a new year, some of the sound types we'd identified were gone, and some new ones were present. Looking at recordings from over 14 years, we discovered that none of the sound types present in the first year were evident 10 years later.

Lou was skeptical of our finding, and it made no sense to me either. Very few adult animals will abandon all the familiar sounds they use regularly and switch to using a whole new set of sounds. Most mammals do not drop or add any sounds to their vocal repertoire after reaching maturity. Such behavior seemed even more improbable given that singers are often a kilometer or more away from other whales when they sing. Animals that use sound to communicate over long distances almost always use a fixed set of sounds. That way, if the sound is faint or distorted after transmission, the listener can still potentially guess at the original sound. Changing communication signals across years makes the perceptually challenging task of communicating over long distances even more difficult.

It was after contemplating this conundrum for several months that I experienced my first scientific epiphany. Epiphanies are rare and not always right. In science the ones that stick are canonized as turning points. Falling fruit causes Isaac Newton to suddenly (supposedly) realize that astronomical entities and more mundane objects all obey the same laws of motion. Otto Loewi dreams of persistently beating hearts in disemboweled frogs, jumps from bed, and rushes to his lab to confirm that chemicals are the medium through which nerve cells transmit information. And, most famously, the speedy streaker Archimedes leaps from his bath to proclaim his discovery of how to identify pure gold. Scientists are constantly seeking for puzzles to solve wherever they can find them. Scientific epiphanies are the progeny of curiosity and consternation.

Unlike Archimedes, my new idea did not arise while I was chilling in the bath. Instead, I was reading a book about biosonar in dolphins, written by one of the dolphin experts that drew me to Hawai'i in the first place, Whitlow (Whit) Au. Whit published *The Sonar of Dolphins*[13] in 1993, the year I started grad school. One chapter describing some intriguing experiments on echolocating belugas caught my attention. In these experiments, targets were moved farther and farther away from the animals to determine the maximum distance at which belugas could perceive targets using echolocation. To the experimenters' surprise, when targets

---

13. Even now, *The Sonar of Dolphins* contains the most extensive scientific assessment of dolphin echolocation available.

were positioned quite far from the belugas, the belugas switched from producing the standard series of echolocation clicks that many cetaceans favor and began producing clicks in short, machine gun–like bursts. It was in thinking about why belugas would do such a thing that the gears in my head "clicked."[14] I realized there might be advantages to using different modes of echolocating when targets are quite far away. I wondered if the songs of humpback whales were a different kind of sonar signal, one specialized for detecting targets from vast distances. Maybe, like belugas, humpback whales had evolved some sonic strategies that helped them echoically perceive objects from distances far beyond what scientists imagined was possible.

My first reaction to this idea was incredulity. I did not know why it was wrong, but it seemed clear to me that it must be. I felt something akin to embarrassment that I couldn't figure out why my weird idea was physically a nonstarter. I spent a few weeks poring over various papers and equations trying to find the physical facts that would rule out the possibility that humpback whales could use their song sounds as sonar, but with little luck. Ultimately, I decided that my understanding of underwater acoustics was too limited for me to discover what it was that I was missing. It was time to reach out to the experts. I emailed two scientists that I was confident would be able to set me straight: Whit, the dolphin sonar expert, and Neil Frazer, a professor of geophysics and an ocean acoustics guru whom I had once briefly met in a group discussion related to dolphin sonar research. My message laid out the facts I had collected about how I thought humpback whales' song sounds should propagate over long distances, how they might reflect from underwater objects, and how it seemed, in principle,

---

14. The role of serendipity is often overlooked in discussions of scientific epiphanies. Isaac Newton gets the adulation for his momentous insight about gravity, but what about the timing of that apocryphal apple fall? Why doesn't the apple tree get some credit?

like they might be useful as a long-distance sonar signal. The message ended with a sort of digital shoulder shrug and a request for feedback about what it was I was missing.

Though they knew me only in passing, both scientists sent back thoughtful replies. Whit answered that humpback whale songs could not possibly be used for sonar because the distances were too far, making any echoes undetectable. Neil's reply was almost the opposite of Whit's. "Of course!" he responded. "Humpback whales must be using their songs for sonar!" The fact that a physicist specializing in ocean acoustics judged the sonar hypothesis not only physically plausible but even likely encouraged me to consider it more seriously. Whit's skepticism was a bit of a downer, but when I mentioned it to Neil, his take was that no one really knew how far a humpback whale might be able to echolocate because no one had done any experiments to test their limits or even estimated what those limits might be using acoustic simulations. This missing knowledge made me even more curious about what singing humpback whales might be able to perceive using sound.

I began rereading many of the papers on humpback whale singing behavior but now mentally replaced the word "singing" with "echolocating" to see whether the observed behaviors of whales still made sense. Not only did the reported observations make sense, but in several cases, observed whales' interactions with, and reactions to, singers seemed more understandable. For instance, the sonar hypothesis explained why whales change their swimming trajectories from long distances when avoiding singers. If singers can locate other whales only when those whales make sounds, then there is no need for other whales to engage in evasive maneuvers. The reluctant joinee could simply go silent and swim anywhere other than toward the singer, and the singer should have no way of detecting its position (like a cheating Marco-Polo player). If singers can perceive other whales echoically, however, then it's a quite different game of chase. Specifically, if singers can localize silent whales from long distances using echolocation, then staying quiet is less likely to be an effective avoidance strategy.

The sonar hypothesis also explained other puzzling humpback whale behaviors observed by scientists. For instance, sometimes singers suddenly stop singing and then swim rapidly toward whales located many kilometers away. Perhaps the singer detected the location of these distant whales echoically. Singers have also been observed to intermittently start singing again when attempting to join a whale that is trying to evade them. If singers are echolocating, this would help them reassess where the avoidant whale is heading.

These past observations confirmed a surprising prediction of the sonar hypothesis. If singing humpback whales are detecting targets at long distances using echolocation, then they should sometimes act like they are attempting to locate distant targets when singing. Additionally, singers should sometimes act like they have detected a distant (silent) target while singing and change their actions based on this new knowledge. Discovering that singing humpback whales do sometimes act like they are "seeing with sound" helped me to build up the courage to approach Lou with this seemingly preposterous idea. I knew that if I simply told him the idea, he would likely dismiss it out of hand, so instead I set up a meeting with him that was about "why humpbacks sing."

Why the roundabout reveal? When I first arrived at Kewalo Basin, I had spent months writing up a proposal for an experiment looking at whether dolphins could learn number concepts. It was literally hundreds of hours of work. After I gave the proposal to Lou, it sat in a pile on his desk for months. Finally, at an unrelated meeting on some logistical topic I asked him what he thought about my idea. He did not remember me giving him a proposal, so I pointed out its location on his desk. Lou picked it up, looked at the title, and said, "It's too controversial." Lesson learned.

I began the meeting with Lou by asking him questions about the known actions of singing humpback whales and other whales in their vicinity, which he of course could easily answer. I asked the questions in a particular sequence so they became increasingly difficult to explain in terms of standard hypotheses about why humpbacks sing—for example, "What happens when you broadcast a recording of a singer to a female humpback whale?," and "What sorts of things do singers typically do when they finish singing?" About thirty minutes into the meeting, Lou stopped, slowly smirked at me, and said, "You think they're echolocating." Then the interrogation switched sides, and he began pelting me with a slew of questions: "Why are they only singing on the breeding grounds?" "Why aren't the females singing?" "Why are they changing their songs each year?" and so on. I was prepared for many of his questions because I'd thought about them a lot. But for some I could only respond, "I don't know." Arriving at the idea himself clearly made him more open to thinking through it. I will not say he was convinced at the end of the meeting that humpback whale songs were sonar signals,[15] but at least he was not hostile to the idea. Considering his extensive experience studying humpback whale behavior, including the behavior of singers, I considered that a major victory.

Over the next year, I worked with both Neil and Lou to further explore the possibility that singing humpback whales might use some or all of the sounds within their songs to monitor unseen events at long distances. We compared the known sonar signals of echolocating bats and cetaceans with the sounds that singers used. We extensively analyzed the behavior of singers and surrounding whales. We conducted computational analyses to determine the possible ranges at which singers might be able to detect whale-sized targets, given the physical properties of those targets,

---

15. Neither was I.

the ocean environment where whales were singing, and the reflective properties of whales. I summarized our results in an abstract accepted for presentation at a scientific conference on acoustics, conveniently to be held in Honolulu.[16] My first official contribution to science!

By the time the conference finally started, I had been developing and practicing my presentation for months. When it came time for me to present, I was surprised to see that the room was packed, with many attendees standing against the walls because of filled seats. I nervously went through my presentation, managing to calm myself down somewhat by the middle and to complete it just before the allocated time ran out. The audience politely applauded, after which the moderator announced that there was time for a single question before the next talk.

A man stood up near the back of the room, whom I recognized as a top marine mammal researcher from California. "You are ignoring 50 years of behavioral and ecological research!!" he yelled and then ranted for the next 4 minutes about how everything that I had presented was pseudoscientific garbage. By the time he finished venting, everyone in the room was staring at me as if they expected me to recant the entire presentation. Instead, I said, "I'm not sure what your question was, but all the information I presented was based on established facts, and those facts are consistent with the possibility that humpback whales are using songs as a kind of sonar." Then I left the podium. When I exited the meeting room, two whale song researchers tag-teamed me in the hall and began listing all the reasons why humpback whale song could not be sonar, as well as ways in which my reasoning was incoherent. I attempted to respond to their concerns, but that only seemed to make them more irritated. I was happy to concede that I had not "proven" that singing humpback whales were echolocating but

---

16. A mecca for scientific conferences.

remained unconvinced that any of the reasons they listed was good enough to dismiss the possibility.

Why all the drama in response to a grad student giving a talk at a conference? To the experts I was not only criticizing a sacred cow of the marine mammal field by implying that humpback whales were not really singing but also suggesting that they and their colleagues had made a major scientific blunder. From their perspective I was not evaluating a novel scientific hypothesis; I was attacking them and their work.

Although I viewed the conference presentation as a step forward, Lou viewed its outcome as a red flag. Afterward, he disassociated himself from any further discussion of the sonar hypothesis and declined being included as an author on any paper discussing it. He did not deter me from pursuing the idea, however, and ultimately approved including it as part of my dissertation project. Neil, in contrast, viewed the conference brouhaha as a positive sign, noting that many novel scientific ideas are at first aggressively dismissed.

Pragmatically, however, this resulted in whale song experts savaging any scientific paper I submitted discussing the topic, meaning that no scientific journals would publish my ideas or findings.[17] In the world of university careers, no publications means no job offers. So I switched gears, left whale bioacoustics in the hands of the experts, and ultimately got a position in a computational neuroscience lab studying how brains make learning and memory possible—no whales included. At least that was the plan. But, like singers surfacing intermittently to breathe, the songs of whales

---

17. On how articles in scientific journals get published: To make it to print in a reputable journal, every paper must be evaluated by at least three other scientists with expertise in the field. This can be a convoluted and contentious process! First, an editor must decide on which scientists to recruit to read and judge the paper. The scientists who agree to be judges often do not agree on a verdict. It takes only one prominent expert to prevent a scientific paper from being published.

Whale Song Acoustics: Songs as Music

The correct answer to the question, "Why do whales sing?" is "What counts as singing?" A singer is "one that sings," and singing is the act of producing songs. Without some way of identifying what counts as a song, it is meaningless to describe whales as singers.

The songs produced by human singers are a kind of music, so it's tempting to think of whale songs as aquatic music. Human singers are musicians who use their voices as instruments to create songs. It's easy to recognize most human songs based on how they sound. But what about songs not created for humans to hear? Are chirping crickets musicians? Does singing with your legs still count as singing?

While the term *song* is straightforward as a description of a musical piece, it is not so simple when used to describe what animals do with their voices—because there are patterns, and then there are patterns. Orcas (the biggest species of dolphin) produce sounds in patterns, but researchers call what orcas do "calling" and not singing. Humpback whales produce sounds in several different kinds of patterns throughout the year, but only one kind of pattern they make is described as "song."

Human songs often contain lyrics. There's not just a pattern in the notes; there's a pattern in the words that convey the meaning of the song. Because whale songs contain so many different sounds and patterns, many scientists and nonscientists have wondered if perhaps their patterned production is conveying some deeper meaning that is lost on human listeners—maybe some complex communication that would be better understood by advanced alien civilizations.

When whale songs are viewed as a kind of music, they can be described in the same ways as human songs: Each continuous sound is like a word, and some words might be repeated. Some sets of words may also be repeated, as in the song "Row, Row, Row Your Boat," in which phrases can be repeated many times with few ill effects. Through the lens of human music, humpback whale songs are a lot like "Row, Row, Row Your Boat" gone wild. Imagine that each phrase within the song could be repeated a random number of times before moving on to the following phrase and that the singer continually repeats this extended song for 10-plus hours nonstop.

Musical notation shows each sound sung by a singer as an oval.

Blow Blow Blow Your Note

Folk song

Blow, blow, blow your note Drift-ing as you dream.

Heedfull-y, heedfull-y, heedfull-y, heedfull-y, song is but a stream.

The higher the oval, the higher the pitch of the sound.

Songs can be divided into "phrases" based on what seems to go together.

Musical songs are often represented as notes and words on a musical staff.

have repeatedly reared their monstrous heads throughout my career. I may have been done with singing humpback whales post Hawai'i, but they were not done with me.

Echoes of the conceptual chain reaction sparked by my mercenary work with IBM engineers at Kewalo Basin continue to reverberate far beyond what anyone might have reasonably expected, culminating in the book you're reading now. Through contemplating the inner world of whales, I gained a new and deeper appreciation of ocean acoustics, auditory physiology, animal behavior, scientific politics, and brain plasticity. I also saw firsthand how seemingly unconnected threads of ideas, evidence, and mundane daily events can converge to make new scientific concepts possible.

This book is part persuasive essay, part scientific tutorial, and part party-pooping polemic. Each chapter (starting with this one) features cross-species comparisons in art and boxes. If you peruse only the art, you will learn a lot. If boxes are more your thing, then you'll come to know birds, sheep, and songs a bit better. On that note, the box on pages 24–25 provides a first take on the acoustics of whale songs that considers how musical they might be.

In chapter 2 you will learn more about what life is like for singing whales. Philosophers have pondered whether it's possible that you are a brain suspended in a vat, hooked up to a computer that simulates all the things you believe you are experiencing.[18] Some have dismissed this scenario as absurd. But a whale singing at night is literally a brain floating in a vast vat in which the only significant sources of stimulation are sounds coming in from all directions. When whales sing, they are constantly constructing their worlds through their ears.

_____

18. Yes, this is also the basic premise of the *Matrix* movie franchise.

CHAPTER 2

When Whales Sing

WHALES SING ALL THE TIME. Whales in the Southern and Northern Hemispheres are somewhat out of sync, but at any given moment, some whales are always singing somewhere. It took millennia for humans to realize this, though, because the world that singing whales experience is so foreign to us. A singing whale's world closely approximates a sensory deprivation chamber—a place where one weightlessly floats, absent any meaningful visual or tactual sensations. When faced with this "unnatural" scenario, the human mind tends to go off the rails. Neuroscientist John Lilly spearheaded the earliest investigations of peoples' reactions to sensory deprivation.[1] Lilly conducted experiments on himself while floating within a sensory deprivation chamber of his own design. There, he discovered that depriving one's senses led to all kinds of unique perceptual and spiritual experiences. What then might a singing whale be experiencing as it floats in the ocean darkness?

---

1. John Lilly also indirectly assisted in the discovery of humpback whale song because he trained Scott McVay in the analysis of dolphin vocalizations. It was this training that led Roger Payne to seek Scott's assistance in analyzing those early recordings—Scott McVay was the English major mentioned in chapter 1 who with his wife first observed the repeating patterns that are now called humpback whale songs. John also constructed a semi-aquatic house in which humans and dolphins made out.

Left alone in a sensory void, humans create their own imaginary worlds. Singing whales don't have that luxury. Sensorially impoverished states are the foundation of their natural experience. Like a human entombed in a sensory deprivation chamber, singing whales create a perceptual world from mere scraps of sensory inputs. But, unlike humans, the worlds that singers construct must match the physical reality of their ever-shifting ocean habitats. Singing whales must transform the lightless oceanic void into a mentally and physically explorable seascape.

This is perhaps the main difference between investigating whale songs from a psychological versus a biological perspective. Whale biologists are fixated on figuring out how singing males get females pregnant, whereas whale psychologists[2] are more interested in understanding what's on a singer's mind when it sings. Are singers mindlessly reacting to hormonally driven mating impulses? Or are they actively constructing perceptual experiences that will enable them to select future actions more effectively? Deciding between such alternatives requires a closer look at the timing and circumstances of song production across years, months, days, and hours.

Winter Wonder Coast

Humpback whales have been around for about 5 million years and are presumed to have been singing throughout that time. Human mariners, in contrast, have existed less than a million years, with the first anecdotal reports of singing whales appearing a mere 200 years ago in the logs of ship captains. And, as noted in chapter 1, scientists did not collect definitive evidence of whales singing until about 50 years ago. Some whale experts were still claiming that the larger whales made no sounds at all as recently as the 1950s!

---

2. If anyone could reasonably be called that.

Whale *bioacoustics*—the scientific study of sound production, reception, and usage by whales—is still in its infancy.

Part of the reason whale bioacoustics took so long to get going is that humans are not great at hearing sounds underwater. Humans also tend to be disinterested in things they don't naturally sense well. Arguably, the only reason that scientists became aware that whales sing is because of humans' excessive interest in killing each other. That is, the technological imperative to find new and more effective ways to kill people, and avoid being killed by people, ultimately led to the discovery and study of whale songs.

Hydrophones—microphones designed to work underwater—were developed in the early 1900s, enabling people to detect artificial sounds underwater. This sensory prosthetic made the detection and sinking of submarines possible, with the first hydrophone-mediated kills happening in 1916 during World War I. From then on, hydrophones became the primary means of detecting submerged vehicles.

The invention of hydrophones also led to the development of sonar systems. The acronym "SONAR" (SOund Navigation And Ranging) was created to mimic the acronym "RADAR" (RAdio Detection And Ranging) because of similarities in how the two technologies use reflections to localize targets.[3] Military sonar, which involves using underwater speakers to broadcast sounds and hydrophones to detect the echoes those sounds produce, was used extensively to attack submarines in World War II.

These technological advances in human-made sonar coincided with the first demonstration of bats using self-generated echoes

---

3. Both radar and sonar begin by sending waves out into the world. These waves bounce off objects in the environment, and their reflections are picked up by multiple sensors. The times that returning waves arrive at different sensors reveal where the reflections came from. The intensity of the waves reveals the sizes of the reflectors.

to navigate during flight in complete darkness.[4] Donald Griffin, the scientist who conducted these experiments, took advantage of emerging technologies that made it possible to record sounds much higher than what humans can hear (*ultrasound*). He recorded bats flying in darkness and found that they were producing streams of inaudible sounds. He then established through a series of tests that bats were using echoes from these sounds to navigate in flight, a phenomenon he called *echolocation*. Echolocation is sometimes called biosonar in reference to military sonar because the same principles apply to both. In the context of animal bioacoustics, the terms "sonar," "biosonar," and "echolocation" are effectively synonyms.

About a decade after World War II ended, the US Navy began using long chains of hydrophones (called *arrays*) on the seafloor to continuously monitor locations where foreign submarines would be a major risk. Operators listening to signals from these hydrophone arrays certainly heard a variety of whale sounds, including songs, though they had no way of knowing for sure what or who was responsible for them. One such operator, intrigued by the non-submarine sounds he heard through an array installed near Bermuda, decided to investigate their source. In the 1950s he built his own hydrophones and connected them to a buoy anchored to the seafloor. By monitoring what was happening around the buoy, the operator became convinced that humpback whales were the source of the most intriguing and complex sounds. More than a decade later, he told Roger Payne about his observations and gave him tapes of the sounds he had recorded—the same recordings that would later reveal the repetitive structure of humpback whale songs through printouts spread across a living-room floor.

---

4. The Italian scientist Lazaro Spallanzani had discovered that bats could use inaudible sounds to navigate 140 years earlier, but most scientists discounted his findings.

Roger Payne is often credited with introducing the scientific world to the idea of whale song in the 1970s. Roger was not the first researcher to describe the sounds of whales, however. As early as 1952, recordings collected off the coast of Hawai'i presumed to be recordings of whales were described as "having a rather musical quality." The researchers also noted that these early recordings varied seasonally, peaking during the spring months. The Winns, those whale song pioneers introduced in chapter 1, also found that whales mainly sang during late winter and early spring, the time when humpback whales are most likely to conceive.

So some of the first observations gathered about singing humpback whales suggested that they sang more and were more audible in tropical "breeding areas" (such as near Hawai'i and the West Indies) at times when breeding took place. These data caused many scientists to associate singing with breeding. Importantly, however, the tropics are not the only places that humpback whales hang out. Humpback whales spend much of the year in nontropical locations, either feeding or traveling between feeding and breeding grounds. Consequently, one reason that researchers heard singing whales most often during the late winter and early spring is that whales were present in these areas when scientists were attempting to record them. Recordings from other habitats were scarce. Nevertheless, many researchers concluded early on that humpback whales sang only (or mainly) when they were attempting to mate. In this way, *when* humpback whales sang became tightly coupled to explanations for *why* humpback whales sang—correlations between singing and mating were used to infer a causal relationship between the two.

This again is a case of first impressions having big impacts on later research directions and interpretations. Recall from chapter 1 that no scientists thought of any cetacean vocalizations as songs until the Paynes described humpback whales as singing in the 1970s. The correlation between peak singing and seasonal

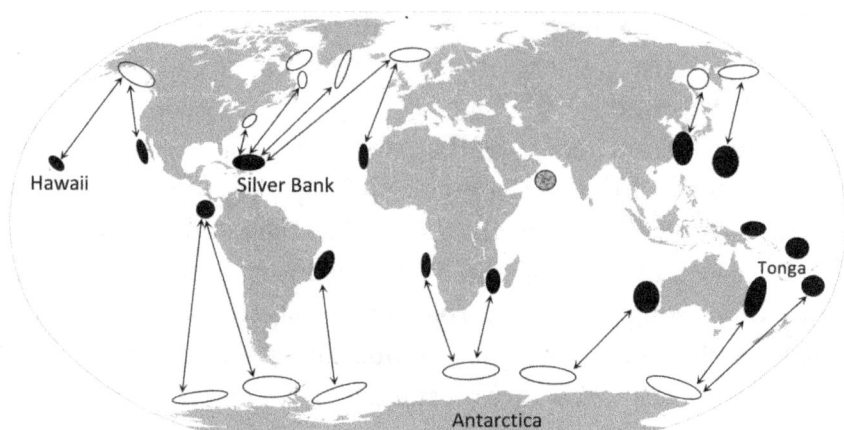

Most humpback whales migrate between polar regions (feeding areas, *shown here as white ovals*) and more equatorial regions (breeding areas, *shown as black ovals*) every year. Different groups of genetically related whales (called *populations* or *stocks*) tend to follow predictable paths each year. Songs have been recorded in every environment that humpback whales inhabit, though their prevalence within specific areas varies seasonally. Only one population of humpbacks doesn't migrate annually (*shown as a splotchy gray oval*).

breeding further cemented the idea that whale song was functionally equivalent to birdsong. From then on, hypotheses about whale song were based almost exclusively on how birds use songs: to attract mates, defend territories, and synchronize reproductive activities.

I feel I should point out here that my goal in this book is to summarize what most scientists believe about singing whales and why whales sing while at the same time highlighting contradictory evidence. For many years it seemed clear that humpback whales sang almost exclusively during the breeding season. As a result, almost all studies of singing whales were conducted in the tropics, further reinforcing scientists' impressions of the reproduction-specific nature of singing behavior.

More recent documentation of humpback whales' singing behavior paints a somewhat different picture, however. It's now known that humpbacks often sing songs as they swim along their migration paths and when they are at the polar feeding grounds. They've even been observed singing while actively foraging! It's true that more whales can be heard singing during periods when they are in tropical breeding areas, but song production extends far beyond the times when females normally conceive.

For a few months during the summer, humpback whales at their feeding grounds don't appear to sing at all. This silent period is more seasonally specific than is singing, suggesting that when whales devote most of their time to feeding, they have more important things to focus on than singing.

You might think that the decreasing correlation between song production and seasonal mating activities would weaken whale researchers' confidence that songs function primarily as mating displays. That is not what happened. Researchers instead concluded that the breeding season must be much longer than previously thought and that males singing on the feeding grounds (and during migration) were either attempting to mate when conception was less likely or to woo individuals they might meet again months later in the tropics.

In other words, whale researchers changed their definition of what counts as a breeding season to preserve its overlap with singing behavior. The "breeding season" came to be viewed as equivalent to "times when humpback whales sing," regardless of what singers and other whales were doing during those times. Even when humpback whales were recorded singing while actively chasing prey, this behavior was still interpreted as a kind of multitasking in which males sexually displayed to other whales (or practiced their acoustic displays) while eating. Long-held conceptions about the nature and functions of songs colored scientists' observations of singing whales—and continue to do so.[5]

Songs are not the only sounds that humpback whales make, either at the feeding grounds or in breeding areas. Humpbacks frequently produce other kinds of structured sound sequences in both regions. For instance, humpback whales were recorded producing repetitive series of "meows" at somewhat regular intervals in the Gulf of Maine. Humpback whales recorded near Australia also produced series of regularly timed repetitive sound patterns, including some that were included within songs recorded in other years.

Despite these repetitive sound sequences meeting all the qualifying criteria for songs, that's not how scientists classify them. Such vocalizations almost certainly would have been described as songs if whales other than humpbacks[6] produced them. The main reasons these other vocal behaviors are not considered songs seem to be as follows: (1) The sounds differ acoustically from the sequences researchers initially described as songs, and (2) some of the sequences are rarely produced in breeding areas. The consensus seems to be that only the longest, most sophisticated sequences produced by humpback whales should be called songs and that other simpler sound sequences can only be called songs if they match most or all the features of these "prototypical" songs.[7]

---

5. This is typical of how research on whale songs has progressed since its first description: Researchers collected new data and then made the data fit into preexisting frameworks for explaining how whales use sound. Over time, explanations that scientists initially offered up as possibilities transformed into dogma that could not be contradicted, regardless of what new evidence was found.

6. Or insects, or frogs.

7. What bioacousticians currently classify as a humpback whale song is based on historical precedents rather than on any objective criteria related to how humpbacks rhythmically produce and repeat sound patterns. For now, I'm reluctantly following this traditional sorting scheme while acknowledging that it is scientifically unjustified and makes no sense.

Humpback whales can live 80 to 90 years and appear to migrate most years, leading to an annual ebb and flow in their singing. Within that yearly cycle, individuals vary their song production in many ways. First, they do not sing songs from birth. It's not known exactly at what age humpback whales begin singing. Most of the whales that scientists have observed singing are at least 8 years old, which is about when young adult whales become sexually mature. We know this because of their body size: Field measurements of singers revealed that only 15% of them were less than 11 meters (36 feet) long, the typical length of a mature humpback whale.

Second, song production typically varies with the time of day, with the most singing occurring at night. But this is not because humpback whales are nocturnal. Humpback whales are highly active during the day. And so are whale researchers; most scientists analyzing whale songs record singers during the day because that's when they can see what they're doing. We only know that humpbacks sing more at night from analyses of recordings collected from autonomous systems placed on the ocean floor and left there for weeks or months.

Importantly, the experience of night for a whale is different from a human's because (1) visibility in the ocean is often low during both night and day, such that seeing is restricted to short ranges for both time periods, and (2) humpbacks are never fully unconscious while sleeping, so there's a less clear divide in states of awareness between night and day. How singing behavior relates to sleep cycles is not known. Singing humpback whales can produce songs nonstop for 20-plus hours, and fin whales are known to do so for more than 40 hours straight. So either whales can sing while semisleeping, or they have no problem singing for days without getting sleepy. For sure, their daily cycles of singing do not map onto those of curious scientists in boats.

SHEEPISH WHALES

You might think, given their fishy form, that the most appropriate species to compare with whales would be animals that look like them, such as whale sharks. If what you're interested in is feeding behavior, then this intuition is correct: Whales and whale sharks both dine on krill, and you would be right to guess that their strategies for collecting these tiny creatures have converged. If, however, your goal is to understand how whales reproduce, then your best bet is to compare them to critters that share more of their DNA. The group that shares the most genes with whales, the artiodactyls, includes hippopotamuses, goats, sheep, and a host of other familiar mammals.

No artiodactyls other than whales are known to sing, which is why scientists rarely draw comparisons between singing whales and their genetically close cousins. The one exception to this is a whale researcher named Jim Darling, who pointed out in the early eighties that "there are some remarkable similarities between male-male and male-female interactions of mountain sheep and humpback whales on the breeding grounds." Chapter 8 will detail the behavioral parallels between mating whales and sheep that Jim noticed. For now, just imagine a bevy of frat boys vying for the attention of a disinterested sorority queen.

Given that mountain sheep neither look nor sound like humpback whales and live in very different environments, it might seem like an odd comparison. But the close genetic relationship and similar mating strategies between mountain sheep and humpback whales make comparisons between these species potentially more informative than comparisons between whales and birds or fish. Biologically and behaviorally, sheep and whales are a closer match than birds and whales—except for the fact that sheep don't sing.

So why don't mountain sheep sing? According to the sonar hypothesis, the main function of singing is searching, so if mountain sheep had an alternative means of searching, they would gain no advantage from singing. And indeed, mountain sheep do have an alternative means of searching, which is by using their eyes. Interestingly, when male mountain sheep (rams) visually search for potential mates (ewes), they are often initially alone, stationary, and positioned on high peaks. Extended attentive staring allows them to potentially detect the locations and movements of distant females entering breeding areas. This kind of vigilant state should sound somewhat familiar.

Behaviorally, the most notable feature of singers is that they are usually alone. The earliest reports of singing humpback whales noted that all the singers recorded were unaccompanied. Later studies over a longer period found that about 85% of singers recorded during the day were alone. Most of these studies were conducted in tropical areas, so it's possible that singing in other regions, or at night, is less lonely.

Singers, whether solitary or accompanied, tend to be more stationary than nonsinging whales and may hang out in the same spot

Overleaf: *Searching from afar.*
Rams visually scan large areas from high vantage points when searching for ewes during the breeding season. Singing humpback whales also search for prospective mates but have to use their ears rather than their eyes to find them.

for many hours. Off the coast of Brazil, lone singers in shallow water sometimes anchor their heads on the seafloor such that their tails stick up above the surface while they sing. When humpback whales sing, this is often their primary or only activity, rather than an action that occurs in parallel with other mating strategies or with courtship movements like those performed by most songbirds.

In summary, humpback whales often sing when they're old enough, when it's dark, when little else is happening, and when they're alone. Males consistently sing in locations and contexts where a reproductively receptive female might show up. Like rams, male singers cannot predict exactly when or where a female will show up, whether any female that does appear will be receptive to mating, or when competing suitors might complicate matters. Most importantly, when humpback whales sing, they're often doing so in situations where few sensations other than sounds can provide them with any relevant information about what's happening around them. A stationary singer, floating alone, submerged, is probably as close to a disembodied brain/mind as has ever existed on Earth. Thus, to understand why whales sing when they do, it's critical to closely consider what these oversized floating brains are like.

The Mind of the Humpback Whale

When as a naive engineering student I first began thinking about how one might make computer hardware more like brain wetware, my first inclination was to look at baby brains. I knew that many people had already tried to make computers function more like brains, to no avail. I guessed, or hoped, that this lack of progress was because engineers were aiming too high by trying to recreate the kinds of fancy processing that happens in adults' brains. By my thinking, baby brains should be performing much simpler functions that might be easier to simulate with silicon. It didn't take me long to realize that scientists knew less about how babies' brains worked than they did about adult brains, making simulating them out of the question.[8]

This initial dead end led me to explore brain function in other animals to try to get a better handle on what simpler (I presumed) brains could accomplish. To my surprise, despite decades of research on nonhuman animals, there seemed to be few known principles that enabled one to predict what any given animal might be able to do based on its neural resources. Although there were lots of discussions about brain size making a difference, the scientific literature on this topic had little to say about how one might predict what 2,000 neurons could do that 1,000 neurons could not, or about the functional consequences of having fewer or more neurons of different shapes and sizes. Most intriguing to me was the possibility that animals like cetaceans, who have quite large brains,[9] might possess capacities beyond what most neuroscientists assumed.

Interest in the exceptional brains of dolphins and whales picked up steam in the 1930s. A neuroanatomist named Orthello Langworthy from Johns Hopkins University was perhaps the first scientist to describe the brain of a dolphin as "truly remarkable" and to suggest that studies of cetacean brains could have "wide and important significance." He noted that the folds within a dolphin's brain were extremely complex, even more convoluted than what is found in humans' notably wrinkly brains,[10] and that their auditory nerves were enormous.

---

8. Mainly because it was/is extremely difficult to observe what's happening in a baby brain or to tell functional activity from not so functional activity.

9. Including the largest brain on the planet, which belongs to the sperm whale and can weigh as much as 20 pounds.

10. The "wrinkles" or folds visible on the surface of some brains are what happens when nature tries to squish an overly large cerebral cortex into a confined space—more complex folds mean an animal's cerebral cortex is quite large.

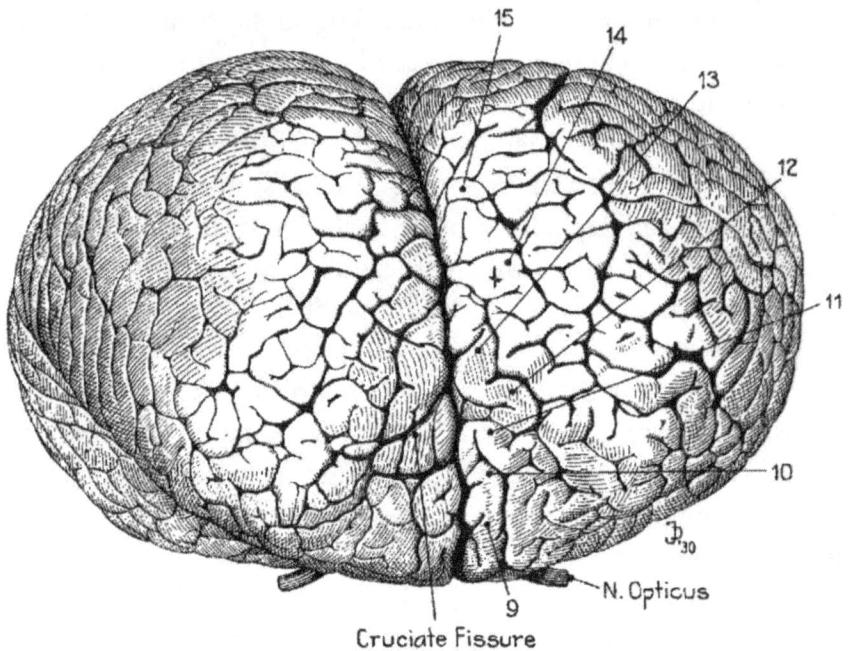

15 14
13
12
11
10
ᴣ₃₀
N. Opticus
9
Cruciate Fissure

Orthello's early sketch of a dolphin's brain, detailing the complex folds that were evident in the large cerebral hemispheres. *Source*: Reprinted with permission from Langworthy (1932).

These early investigations showed that although the brains of cetaceans differ greatly from those of mammals that live on land,[11] they are clearly mammalian brains, with all the same basic anatomical components. Orthello described the dolphin brain as similar to that of a giant hedgehog brain because of the neurons' organization within the dolphin's cerebral cortex.[12] He provided no explanation for why the dolphin's hemispheres were so complexly convoluted or why its auditory circuits were so gigantic relative to other mammals.

---

11. I'll describe some of these differences in more detail in chapter 9.
12. The wrinkly parts that cover the surface of the cerebral hemispheres.

Two major proponents of the idea that cetacean brains might be capable of more than your average hedgehog brain were neuroscientist John Lilly[13] and biologist Karl-Erik Fichtelius. Both independently reached this conclusion in the late 1960s, and both laid out their arguments in books written for a general audience, books that I stumbled across while searching for information on "simpler" animal brains.

In *Man and Dolphin*, Lilly (1961) proposed that when a brain reaches a certain size, it gains the capacity to learn language. He hypothesized that since a dolphin's brain is comparable in size to that of a human, dolphins should be able to learn a language of their own.[14] John also proposed that dolphins might be able to learn a simplified human language, making interspecies conversations possible. Fichtelius and Sverre Sjölander, in the provocatively titled book *Smarter than Man?*, suggested that the enormous brains of sperm whales gave them enhanced listening and communication abilities. They pointed out that sperm whales might engage in intellectual activities different from those favored by humans and that it was scientifically unjustified to assume that human mentality is either the apex of what is possible or the best-case scenario.[15]

Most neuroscientists and biologists discounted these sorts of speculative ideas. Despite my neuroscientific naivete at the time I discovered these books, I couldn't help noticing that the naysayers' counterarguments were driven more by skepticism about animals' mental capacities than by any solid theory of how variations in brain circuitry affect an organism's mental endowments or cognitive potential.

Chapter 9 will explore in more detail the critical issue of what cetaceans' brains can do. For now, you just need to know that

---

13. Yes, the same guy who spent his days lounging around in a sensory deprivation chamber.
14. What some have called delphinidese.
15. Noting many of human societies' perennial woes as a case in point.

singing whales, like most other cetaceans, have considerable neural-processing resources at their disposal and that they definitely use their brainpower in ways that you do not.

For starters, singing whales are always conscious. Cetaceans are unihemispheric sleepers, meaning that one-half of their brain remains in an awake state while the other half is on hiatus. Presumably, whales and dolphins never dream because if they did, one-half of their brain would be experiencing internally generated scenes while the other half witnessed the external world.[16] This surprising mode of mammalian consciousness was discovered by none other than John Lilly when he accidentally killed several dolphins by anesthetizing them—unconscious dolphins don't breathe. A sleep-free life is already hard to imagine, but even more impressively, dolphins[17] appear to be free from mental fatigue. In a relatively recent study in which a dolphin was trained to detect the presence or absence of specific targets using echolocation, the dolphin was able to perform this task nearly flawlessly, nonstop, for over two weeks. At a minimum, singing whales' impressive brains likely provide them with similarly extraordinary awareness and vigilance.

What are whales doing with all that excess attentional capacity? I noted earlier that when humpback whales sing, they usually are alone. Physically alone, that is. Whale songs travel many miles so that, in principle, singing whales could stay mentally connected to each other over vast distances. And it is often the case that in areas where whales sing, more than one singer can be heard. When multiple singing whales are audible, they are said to be chorusing.[18]

---

16. Effectively creating two coexisting but inconsistent perceptual states.

17. And likely other cetaceans as well.

18. Levels of sound production within a chorus vary based on the number of singers, so scientists can use recorded choruses to estimate how many singers were part of the sonic party.

Whale Song Acoustics: Songs as Chorusing

Like singing humans, singing whales can vocalize on their own or in places where multiple whales can be heard. When several whales are audible, scientists say that the whales are chorusing. This description is a bit misleading, however, because the singers within whale choruses are never close together but typically spaced out across several kilometers. It's only because the songs are so intense and propagate so far that the whales appear to be singing "together." Also, unlike human choruses, individual singers don't seem to coordinate the timing of their vocalizations in ways that would produce sonorous rhythms or harmonies. Humpback whale choruses are cacophonous and early on were referred to as a barnyard chorus. Consequently, it's often assumed that singing whales are doing their own thing without paying much attention to the specific "notes" and timing of other whales' songs.

The fact that singers are not collectively keeping time when they sing in groups doesn't mean they're ignoring each other. In some cases, singing whales seem like they might be matching parts of their songs to those they hear other whales singing. A singing whale may also adjust the timing of its song so it's not singing the exact same thing as surrounding singers at the same time, a bit like singing "Row, Row, Row Your Boat" in a round.

Multiple whales can often be heard singing in choruses in breeding areas, but whether they constitute a singing community is unclear. Singers clearly aren't vocally cooperating like members of a church choir. But they may choose which sounds to produce based on what other whales in the area are doing.

Squiggly up-arrows mean pitches should rapidly rise.

Co-occurring sounds are vertically aligned.

Blow Blow Blow Your Note (Remix)

This means be silent for a fixed duration between sounds.

The songs produced by chorusing humans are usually represented as expanded versions of simple songs, with multiple notes presented in parallel. Chorusing whales, in contrast, do not coordinate the timing and harmonies of the songs they produce.

Regardless of whether songs are used as sexual displays or sonar signals, whales singing within a chorus[19] must be able to selectively focus on individual singers within the din. If singers use echoes from songs to monitor their surroundings, as the sonar hypothesis proposes, then doing so within a chorus is likely to be especially challenging since each singer must listen for the specific echoes they generated. The capacity to vigilantly monitor the surrounding songscape becomes critical in this context.

According to the sonar hypothesis, when humpback whales sing for hours on end, they effectively create acoustic "webs" in space

---

19. And any listeners attempting to gain information about individual singers within that chorus.

and time.[20] To detect when other whales have entered their personal space, singers must differentiate small changes in incoming echoes from among the background noise of other whales' songs and the echoes generated by those songs. This is a cognitive feat that humans cannot match, even with technological assistance. When whales sing in choruses, they're engaged in an auditory task much more challenging than the one singers face within a human choir, where everyone's actions are synchronized in time and coordinated in pitch. Humpback whales and other cetaceans have gargantuan auditory circuits in their brains for a reason. When whales combine their massive hearing engines with an apparently unlimited capacity to monitor acoustic events, there is every reason to expect astounding auditory abilities will emerge.

At this point you may be thinking, If singing makes it possible for an individual to continuously monitor its surroundings, then why aren't all cetaceans singing all the time? Chapter 3 will explore the various reasons why not all whales that sing produce the same kinds of songs and why some cetaceans, like dolphins, don't seem to sing at all.

---

20. The auditory analogues of spiders' webs.

CHAPTER 3

Which Whales Sing

BIOLOGISTS BELIEVE SONGS PLAY an important role in whales' love lives for two main reasons. The first is the seasonal coincidence of peak singing and peak conception. This correlation doesn't guarantee that the two behaviors are directly related, however. Seasonal peaks in sound production by echolocating bats also coincide with peaks in conception rates, but no bat scientists are claiming that bats evolved sonar so they could make babies more efficiently.[1] The second, and probably more compelling, line of evidence that whale songs are elaborate mating displays is that almost all singers sexed so far have been males. The question I get most often whenever I talk about the sonar hypothesis at scientific meetings— which I still do occasionally, despite my early drubbing as a graduate student—is "If song is a kind of sonar, then why do only males sing?" Usually, the question is asked in a way that is more of a critique than a question.

If only I could receive a grant every time someone asks me that question, I could pay a team of people to answer it for me. The

---

1. For bats, this correlation occurs because many bats hibernate during part of the year, presenting a very different scenario from whales. Still, the point is that seasonal trends can arise for multiple reasons.

response I always want to give is "What? Only males sing?" because clearly I'm a special kind of clueless to not understand that this one incontrovertible fact discounts any possibility that whales use their songs as sonar. Except this "fact" is a bit more controvertible than you might guess from the claims of biologists.

It's certainly true that lots of males sing in the breeding grounds. But inferring that *only* males sing from the data that have been published to date? That's more religion than science. After all, no scientist has definitively observed or recorded humpback whales having sex (other than male-on-male sex). Should we then conclude that humpback whale males never have sex with females? Of course not. All that this lack of evidence means is that scientists have not yet seen everything that whales do.

And it's not even certain that scientists haven't discovered female humpback whales singing. One early published report specifically mentions a case in which a female seemed to be singing. Twice at conferences, a scientist I didn't know sidled up to me as if a drug deal was about to go down and told me, unsolicited, "I've seen a female singing." I almost wanted to tell both confessors, "Your secret is safe with me" but responded instead with "That's cool!" Others have probably thought they might have seen a singing female but either weren't sure of what they saw or just wrote it off as some unnatural oddity.

In any case, I think the data-grounded way to think about this issue is to consider why all the singers that have been sexed so far have been males. I should point out that this is not a massive number of singers. By my count, fewer than 200 individual singing whales have been sexed in published reports over the last 50 years, with almost all identified as males. Most studies that claim to have sexed singers, however, did not use methods guaranteeing that the source of the sounds was the sexed whale, so past interpretations may be skewed toward confirming prior beliefs.

Despite these caveats, there are good reasons to believe that most humpback whales singing in the breeding grounds are males. I'm in total agreement with other whale researchers in thinking that if you come across a whale singing in the tropics, there's a high probability that whale's a male. So why is that?

Big Boys with Big Brains

The short answer to why male humpback whales are the ones singing during breeding periods is that males are highly motivated to find other whales[2] when mating is a possibility. Testosterone, a hormone known to increase males' efforts to mate, peaks in male humpback whales in the middle of the breeding season. Some scientists believe that surges in testosterone trigger males to start singing. Certainly, higher testosterone levels should ramp up males' sex drive. So if singing increases opportunities for mating, then males should be more likely to sing when testosterone levels are high.

No male humpback whale is going to impregnate a female simply by singing, however. Testosterone levels affect many reproduction-related behaviors. For instance, males' motivations to mate may partly determine when they initiate their long migratory journeys. Variations in testosterone probably also impact how males interact with other males, including singing males. These kinds of correlations between randiness and behavioral changes in males are apparent in many mammals. In most cases hormones change males' attempts at mating in ways that differ from what you see happening in females.

Chapter 8 will provide more details about the various actions that male and female humpback whales engage in during the mating season. For now it's enough to know that interested males outnumber females in the tropics. Sexually receptive females are

---

2. Particularly sexually receptive females.

scarce in the areas where most breeding occurs. In most cases, multiple males seek the attention of any sexually receptive female. Also, as with many artiodactyls,[3] intercourse in whales is a relatively short-lived affair. Once a female is no longer sexually receptive, her attractiveness to males, including any that have mated with her, rapidly fades, and the previous partners go their separate ways.

This is the scenario that male humpback whales find themselves in each winter. Most are motivated to mount females. To do so they need to complete four steps: find a female, swim after her, beat out other equally motivated males, and be in the right place when the female becomes sexually receptive. As part of this process, males sing. Singing in some way enhances males' reproductive opportunities. And there's a decent chance it does so by contributing to one or more of these four requisite steps to mating.

For many whale researchers, correlations between the timing of singing and the sexes of the singers provide definitive evidence that humpback whale songs are for mating. Yes, . . . but how? What exactly is it that the males are doing, or attempting to do, when they sing? And does the fact that males sing a lot during the breeding season mean that only males do it? Or does it mean that songs function in ways that are particularly useful for males attempting to mate? We know from mountain sheep and many other mammals that just because males behave in certain ways when they are female focused (like stalk females or butt heads) doesn't mean they're the only ones that do.

So far in this book, the phrase "singing whales" has really meant "singing humpback whales," which for some scientists is a synonym for "singing male humpback whales." Humpbacks are not

---

3. Recall from chapter 2 that artiodactyls are a group of mammals that includes hippopotamuses, goats, sheep, and more. And they are the group of terrestrial mammals that shares the most genes with whales.

SINGLE GRAY RAM SEEKING SEX PARTNERS

Like humpback whales, mountain sheep mate during a limited period, called a rut, that lasts a few months—October through January. And, like humpback whales, sheep shift their locations seasonally between a summer range and a winter range where breeding occurs. About a month before the rut, rams start migrating to the locations where they will mate. Usually, rams travel 12 or more miles to get there—less impressive than the 5,000-plus miles some whales migrate each year (one-way) but still a long walk.

During the rut, not all females are equally receptive to mating attempts. Generally, ewes will come into heat two days every three to four weeks. Consequently, there are advantages for rams that play the field. Rams need to know the locations of different ewes throughout the breeding season to increase their odds of being in the right place at the right time. Rams use multiple strategies in their efforts to locate and gain access to receptive females, including long-range searching from high vantage points and wandering.

Importantly, rams are the main sheep involved in seeking out potential mates—the males are the seekers, and the ewes are the seekees. And rams only fixate on females that are receptive. Once a ram locates and identifies a potential mate, he attempts to monopolize her by keeping other rams away until she is no longer receptive. The largest rams are typically the most successful at this task. When the ewe reaches the last hours of her receptivity, her "protector" immediately abandons her and heads off looking for the next conquest.

The reproductive interactions of mountain sheep are better understood than those of humpback whales because it's much easier to continuously monitor what sheep are doing (even so, what mating mountain sheep are up to at night is not well-known). Given the close genetic relationships between whales and sheep and the known similarities between some of their reproductive strategies, there's some justification for expecting that many of the factors that determine how sheep behave while mating also apply to humpback whales.

During breeding periods, rams use sex-specific strategies to detect and locate potentially receptive females. For instance, ewes are not positioning themselves on high peaks to better detect and view the movements of other sheep. This does not mean that looking intently from high peaks is a visual mating display that rams use to impress ewes or intimidate other rams, or that only rams can see long distances. What it means is that rams and ewes behave differently in mating contexts.

Closely considering how different rams attempt to gain mating opportunities reveals the complex challenges that arise for each male depending on his situational circumstances. The challenges faced by male humpback whales seeking out sexually receptive females are sure to be at least as complex. Like rams, one of the main tasks facing male (but not female) humpback whales is to put themselves in the right place at the right time. To do so, male whales need to monitor the locations of females throughout the breeding period.

the only whales that sing, however. They are just the species that researchers have spent the most time studying. It could be that humpback songs function differently from other whales' songs. Biologically, it's more likely that if different whales sing in similar ways, then they're doing so for similar reasons. This kind of cross-whale extrapolation is like the argument that scientists use to justify comparisons between whales and songbirds. But the case for cross-whale comparisons is much stronger because of their genetic relatedness.

Seven whale species are known to sing: humpback whales (*Megaptera novaeangliae*), bowhead whales (*Balaena mysticetus*), blue whales (*Balaenoptera musculus*), fin whales (*Balaenoptera physalus*), minke whales (*Balaenoptera acutorostrata*), right whales (*Eubalaena japonica*), and Omura's whales (*Balaenoptera omurai*). All seven are big, ranging from the large minke whale (20,000 pounds) to the ultramassive blue whale (200,000+ pounds). In this club, humpback whales are relatively svelte, at 60,000 pounds. All these species are classified as baleen whales because their mouths are filled with baleen—stiff structures that resemble overused combs. Baleen whales inhabit overlapping ocean regions, migrate long distances, favor feeding on similar organisms, and live several decades. All are promiscuous[4] and polygynous[5] and vocalize so loudly that their songs can travel many miles.

The seven singing species have a common cetacean ancestor, a greatest-grandmother, that sauntered through the seas around 30 million years ago. There's currently no way of knowing if this extinct ancestor sang, but it almost certainly possessed the ability to make sounds underwater. Did mega grandma use those sounds for sonar? That's unclear as well. But there's distinct evidence of an evolutionary split between those whale species that retained their teeth and began using specialized clicks to echolocate (called toothed whales) and those whose teeth gave way to baleen. This split happened around 35 million years ago, making it likely that singing whales' common ancestors possessed the right anatomical parts for using sounds as sonar, if they weren't already using sound in this way.

---

4. Meaning they have multiple sexual partners.
5. Describing the fact that males mate with multiple females.

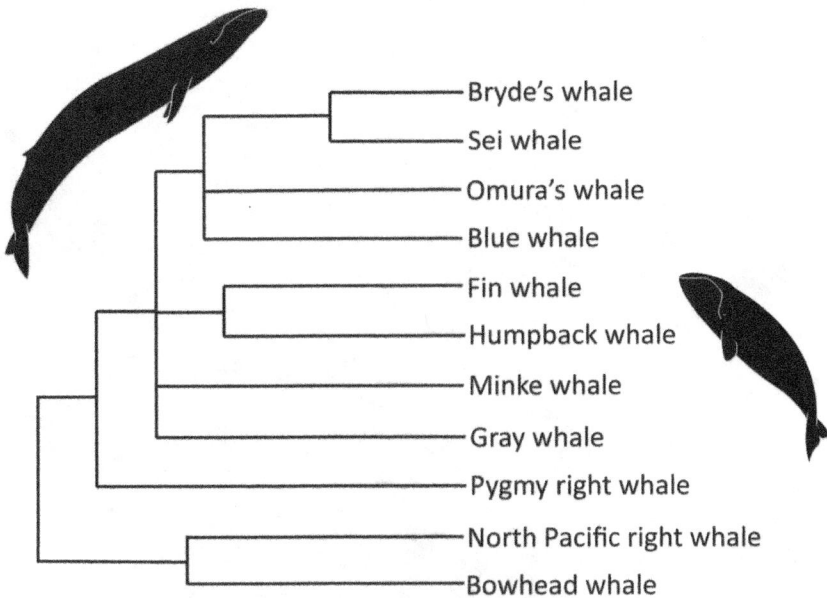

Seven species of whales are known to sing—they are all baleen whales. Gray whales are not known to sing. Bryde's whales, sei whales, and pygmy right whales have not been recorded singing. They are more difficult to monitor, however, so that is not strong evidence that they don't sing. Different species of whales sing different kinds of songs.

Many scientists view the evolutionary split between toothed whales and baleen whales as the point in natural history when some whales evolved into echolocators, and others evolved into singers. Toothed whales clearly evolved along different trajectories from their larger baleen whale cousins, trajectories driven in large part by ancient opportunities to capture and eat specific types of prey. Both groups, however, share many of the adaptations to surviving underwater, including enhanced auditory circuitry and vocal flexibility relative to mammals that live on land.

It's widely claimed that toothed whales are the only cetaceans that echolocate and that they only do so using short-duration clicks. The clicks produced by toothed whales are ultrasonic, meaning

they contain elements that humans cannot hear. Imagine an operatic singer who sings every note higher than the last until you can see her mouth making singing movements but can no longer hear any sounds coming out. Toothed whales can hear these kinds of super high-pitched sounds, making it possible for them to interpret echoes from ultrasonic clicks with high resolution. Some humans have learned to echolocate using clicks, but those clicks are not ultrasonic and can't provide as clear a picture as what toothed whales perceive.[6]

The consensus view on vocal evolution in cetaceans is that those with teeth are the clicking echolocators, and those with baleen are the (sometimes) singing displayers. The tendency of scientists to equate underwater echolocation with ultrasonic clicking is a problem. Many baleen whales make clicking sounds that are sonic.[7] Certain birds and humans can echolocate using audible clicks, so it's a bit ludicrous to assume that cetaceans, animals that have evolved echoic processing abilities way beyond that of any human or bird, aren't able to cope with echoes from lower-frequency sounds. The idea that cetaceans need to use clicks to echolocate is similarly suspect. Military organizations around the world use nonclick sonar signals in the same audible range as sounds produced by baleen whales to monitor and track targets underwater. Sure, it's possible that baleen whales evolved for millions of years without ever

---

6. Chapter 7 will explain why ultrasonic echoes provide higher-resolution images than audible echoes. The short version: The higher the frequency, the smaller the object that will reflect the sound.

7. Meaning they contain frequencies within the range of human hearing.

Opposite page: *Megamouths.*
Keratinous curtains line the jaws of baleen whales, enabling them to strain their tiny prey in bulk. Toothed whales, which include all species of dolphins, sport pointier dentition with which they can wreak havoc on larger fish and mammals.

making use of the behaviorally relevant information provided by sonic echoes, but it does not seem particularly likely.[8]

Some baleen whales don't seem to sing (like gray whales and Sei whales—pronounced "SAY whales"). This might mean that scientists haven't been lucky enough to catch them in the act. Or perhaps the advantages gained from singing aren't relevant for all baleen whales. Another possibility is that scientists discriminate against these supposedly nonsinging baleen whales because their vocal sequences seem less songlike to humans. For instance, gray whales produce a "series of pulses" with acoustic properties similar to those in the songs of right whales and fin whales. The threshold for how regular a series of sounds needs to be to qualify as a song is not based on any criterion other than what researchers have called songs in the past.

So what are researchers calling songs these days? When applied to humans, the term refers to vocally produced musical sounds. It's a bit trickier to say what counts as songs for nonhumans because what sounds musical to an individual of another species[9] might not sound so musical to humans. Historically, people have described animals' sounds as songs if they are melodic and appealing. This is why ducks don't "sing," even though they do produce repetitive series of quacking sounds to attract females. Defining a behavior in terms of its aesthetics makes scientists squeamish, however, and they generally do their best to avoid it.

Recently proposed scientific criteria for calling an animal's vocalizations songs state that the vocalizations need to be complex and learned. These criteria aim to provide unambiguous checkboxes that one can use to decide whether and when any given species sings. Are these criteria unambiguous? Maybe. Let's try

---

8. Interestingly, some of the nonclick sounds humpback whales make while singing do contain ultrasonic elements, but scientists mostly ignore them.

9. If any sounds do.

applying them to one of our singing baleen whales, the blue whale. Is a blue whale's song complex? You can't really tell by listening to a blue whale sing because their songs are infrasonic—too low-pitched for you to hear. Even if you could hear their songs, how would you rate their complexity? What counts as complex enough? You're going to have to decide based on some arbitrarily selected numerical threshold along levels of acoustic complexity or go with what a gaggle of experts thinks is complex enough.

Do blue whales learn their songs? How would you tell? Chapters 8 and 9 will discuss evidence for song learning by humpback whales, but no comparable evidence shows whether blue whales' songs are learned. Some researchers would argue that blue whales probably don't learn their songs because most blue whales living in a particular geographic region produce essentially the same kinds of songs year after year. In this respect, blue whales are a bit like repetitively quacking ducks.

But blue whale songs are surprisingly similar to those produced by chickadees, a species of bird whose songs are even more consistently structured than those of blue whales.[10] Despite this uniformity in songs, experimental studies of chickadees show that their songs are learned! There's no way to run similar experiments with blue whales, so we may never know for sure whether their songs are learned. Do you think the ambiguities noted above have deterred scientists from describing blue whale vocalizations as songs? They haven't so far. It's not clear what, if anything, would change if blue whale researchers capitulated and started calling blue whale sound sequences "sortongs" (sort of songs) instead of songs. Maybe it doesn't really matter what they're called?

Scientists have proposed strict criteria for labeling some vocalizations as songs and others as nonsongs mainly because they want to be able to compare apples to apples when exploring how various

---

10. I'll describe these similarities in more detail in chapter 7.

species communicate. For instance, if I wanted to compare humpback whale songs to songs that I sing in a chorus,[11] it would be helpful if I could apply the same labels and methods to both kinds of vocalizations. Otherwise, my comparisons are going to be more poetry than science.

If scientists want to say anything sensible and defensible about whether whale singing bears any relationship to human singing (or singing by birds), they need to establish that whales are doing something that overlaps in some meaningful way with what other singing animals do. Many musically oriented scientists who study human singing are skeptical of these kinds of human-to-animal comparisons because of the myriad ways in which singing humans use their voices to modulate the emotional and cognitive states of listeners. In contrast, most scientists are comfortable saying that singing whales and birds are doing basically the same things: vocally strutting their stuff to wow the ladies or dissuade the competition.

But are whales and birds vocalizing for the same reasons? Chapters 6 and 8 describe various ways in which humpback whale songs are unlike any known birdsong. Simply saying that vocalizations are songs if they're learned and complex is not going to cut it if your goal is to understand why whales sing. And the goal should not be to understand why male humpback whales sing during breeding periods but why *any* whales sing, in all the ways that they sing, in all the contexts in which they sing. That includes understanding why female whales sing.

Based on the discussion above and any information you've heard from popular sources, you might think that singing by female whales is an extremely rare and abnormal event. You may then be shocked to learn that 100% of the singing bowhead whales that

---

11. I do not.

have been sexed so far have been females. Surprise! Perhaps sing-ing female whales are not as rare as you thought.

Singing is the norm for both sexes in songbirds.[12] There was a time when researchers confidently claimed that only male birds sing, but that rush to judgment came from observing certain spe-cies in restricted situations. If scientists studying animals as acces-sible as birds missed the boat on describing sex differences in song production, how likely do you think it is that scientists studying large, mostly submerged whales have nailed it? If researchers really think that singing humpback whales act like singing birds, then their default assumption should be that both male and females sing since that's the case for most species of songbirds. Whales produce many songs in locations and at times when no researchers are around to sex them. Given the asymmetries in sexed singing whales observed to date, in both males *and* females, it's likely that the two sexes are sometimes singing to achieve different goals.

You now have a better sense of which whales sing, or at least which whales vocalize in ways complex enough to convince scien-tists to describe them as singing. You're also hopefully thinking a bit more about what it means to say that an animal other than a human sings. If so, then you're ready to dive into the issue of whether songs and sonar signals are as radically different as some scientists have made them out to be.

Singing Toothed Whales and Bats?

I noted briefly in chapter 1 that toothed whales produce patterned sound sequences. Some of these sequences appear to be learned, and some are at least as complex as the songs of blue, fin, and right whales. Based on these criteria, toothed whales sing. But no scien-tists describe them as singing. What gives? Why are scientists so

---

12. Surprised again?

stingy when it comes to describing the sound sequences of whales with teeth? Surely, toothed whales can compete with frogs and insects when it comes to vocal skills.

Part of the issue is that researchers studying toothed whale vocalizations are heavily invested in the idea that toothed whales use sounds mainly for echolocation and for communicating during social interactions.[13] Consider the case of sperm whales. Sperm whales sometimes produce clicks in stereotyped, repetitive sequences. Scientists believe that sperm whales use these click sequences for communication while using other less regularly produced clicks for echolocation. Researchers have dubbed the sequences "codas," a musical term that originally referred to singing a repeated syllable at different pitches to end a piece—such as when the Beatles sing "Naaaa naaaa naa na na na naaaa!" in the song "Hey Jude." In other words, when humans produce codas with their voices, everyone agrees they are singing; sperm whales, not so much.

Early reports of sperm whale codas directly compared them to the songs of fin whales and humpback whales because of their similarities. In addition, vocally interacting sperm whales are said to produce duet-like sequences. Oddly, biologists seem to be comfortable using song-related adjectives to describe the patterned click sequences produced by sperm whales but are unwilling to describe sperm whales as singing. My impression is that cetologists[14] have come to view the terms "song" and "reproductive display" as synonymous when applied to animals. From that perspective, if codas are not acoustic peacock tails, then they cannot be songs. Cetacean researchers agree that codas are specialized communication signals and that codas contain the same kinds of clicks that sperm whales use to echolocate. Sperm whale codas are a good example

---

13. But not as a reproductive display.
14. And many other biologists.

of how sounds used for sonar can also be used for communication and maybe even for singing, depending on what one is willing to call a song.

Sperm whales are not the only toothed whales that produce patterned sequences of clicks. Remember those machine-gunning belugas from chapter 1? To recap, when belugas attempt to echoically perceive targets that are quite far away from them, they group their clicks into short packets. These packets are produced in bursts, somewhat like impatient finger drumming. Bottlenose dolphins[15] also use this strategy for echolocating when targets are ultra-far away (think multiple football fields). Scientists commonly describe this kind of rhythmic production of click packets as singing in insects. But since dolphins are making these sound sequences while echolocating, no one describes them as singing. As with sperm whales, dolphins' clackety clicking is not called singing mainly because dolphins don't use the vocalizations as a reproductive display. Belugas and dolphins are excluded from singing with clicks by fiat.

The sounds humpback whales use within their songs are much more varied than the clicks used by echolocating toothed whales. This is why recordings of singing humpback whales sound like singing birds when they're played back at faster speeds. At least one species of toothed whale, however, produces patterned sequences of sounds that are a bit more songlike than rhythmic click trains. Orcas, the largest of the dolphins, produce a wide range of complex vocalizations, including clicks, whistles, and "calls."[16] Sometimes they combine these sounds into predictably patterned sequences. David Bain, a researcher who has studied orca vocalizations both in the wild and in captivity, directly compared orca sound sequences with humpback whale songs in the

---

15. The kind of dolphin you are most likely to see in captivity and in videos.
16. Think screaming baby.

1980s. From these comparisons he concluded that "it is apparent that these two species structure their vocalizations in a similar manner." Pretty much all other scientists ignored his finding, until very recently when a group reported that orcas off the coast of Iceland were consistently producing several "call combination patterns." The researchers described orcas' combos of two or three rhythmically produced sounds as highly stereotyped.

No evidence indicates that orcas use their vocal sequences as reproductive displays. So once again scientists have demurred from describing the orcas as singing. According to the sonar hypothesis, however, humpback whales probably aren't using their sound sequences as reproductive displays either. So it's possible that the similarities between orca call combos and humpback whale songs are not coincidental.[17] Until it's known how orcas and humpback whales use their sound sequences, there's no reason to treat them as radically different kinds of vocalizations. No one has explored the kinds of echoes orcas' vocal sequences generate. In principle, they might also be able to use sonic vocal sequences for long-range echolocation. As of now, scientists assume that orcas only use clicks to echolocate[18] and use all other vocalizations for communication, but we don't really know what orcas perceive.

Toothed whales are not the only mammals that have been excluded from the glee club based on preconceptions about how animals use sounds. Echolocating bats also produce patterned sound sequences like those of humpback whales.[19] For example, a small bat found in Cuba called the little goblin bat[20] switches between

---

17. The vocal sequences these two species produce are not only similar in their stereotypy but also share several acoustic features, some of which are described in chapter 6.

18. Presumably, orcas can use clicks to echolocate over ranges comparable to those documented in bottlenose dolphins (< 1 kilometer), but no experiments actually show how far away orcas can detect targets.

19. But much more rapidly and at higher pitches.

20. Which is almost certainly the model for the monsters in the movie *Gremlins*.

multiple patterned vocal sequences when searching for prey, some of which include rhythmic alternations between two differently pitched sounds (all ultrasonic). Although the patterned sequences produced by little goblin bats and many other bat species contain many of the same structural features evident within humpback whale songs, researchers don't call these sequences songs because bats use them for echolocation. Interestingly, like dolphins and belugas, bats sometimes switch to using short packets of sonar signals when faced with a difficult echolocation task.

Historically, scientists described vocalizing bats in much the same way as toothed whales, believing that bats use ultrasonic sounds (calls, not clicks) for echolocation and sonic sounds for communication. In the last few decades, though, researchers have discovered that some bats do sing. These bats produce songs that are predictably structured, like those of songbirds and whales, and do so in reproductive contexts. Additionally, all the bats that have been sexed while singing have been males, which is probably why scientists feel comfortable calling the vocalizations songs. Some bat species include sounds within their songs that are the same as those for echolocating, providing clear evidence that songs don't need to be radically different from sonar signals to function as communication signals. As with singing birds and whales, most scientists think that singing bats use their songs to deter other male competitors and to attract sexually receptive females.

If bats can use the sounds within their songs to echolocate, might humpback whales also be able to use individual sounds from within their songs to echolocate? In principle, yes. Scientists call the vocalizations that baleen whales make when not singing *calls*.[21] Calls can be distinguished from songs mainly because calls

---

21. Not to be confused with the calls produced by orcas. On second thought, it's fine to confuse them because "calls" is a term scientists have used to describe all kinds of vocalizations.

Whale Song Acoustics: Songs as Sexy Serenades

In humans, singing to secure sex is called serenading. A couple of key things must be true for a serenade to be successful. First, songs produced by different singers must vary in ways that females can detect. If all singers sound the same to a female, there would be no way for her to use the songs to choose between potential mates. Second, according to evolutionary principles of female selection, the songs should have something about them that the females can use to decide which male is "better" (as in likely to make good babies). There must be some properties of songs that females consistently find attractive. So are there such properties within humpback whale songs?

No one has yet identified what might make a whale (or bat) song "hot," though many have tried. Since humpback whales do not produce identical songs, females could potentially find one singer the most alluring. Several complications make this kind of female choice trickier than it sounds, however. For one, individuals do not produce songs identically each time they sing them. So the preferred singer would need to use either some constant "style" or some consistently impressive variation across styles that makes the singer stand apart.

Another complication is that these sonic runway walks are typically heard from long distances. Relatively few singers are physically close to a female. Even in the best-case scenario, a female only has a ringside seat for one singer. If she wants to compare that singer to any others, she has to judge his song against songs that have traveled a mile or more underwater. This kind of long-distance transmission inevitably leads to song distortion, sometimes in the extreme. So picky females are forced to choose mates based on "fuzzy photos" that don't clearly reveal what each singer can produce. A human serenader attempting to woo a female from five blocks away (still much closer than the scenario for whales) is unlikely to succeed. It is questionable whether a singing humpback whale fares any better.

Scientists usually don't describe animal songs with musical notation. Instead, they typically use spectrograms, images like traces of piano key presses as they occur over time.

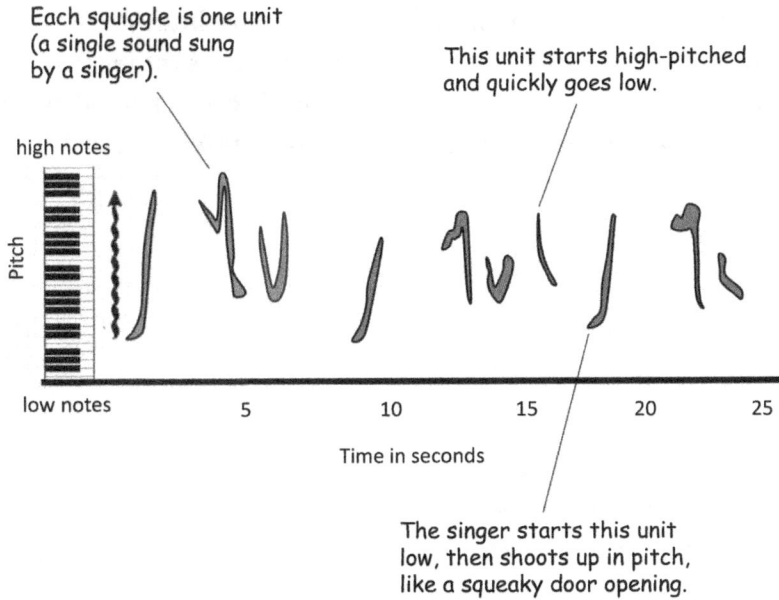

Each squiggle is one unit (a single sound sung by a singer).

This unit starts high-pitched and quickly goes low.

high notes

Pitch

low notes

5   10   15   20   25

Time in seconds

The singer starts this unit low, then shoots up in pitch, like a squeaky door opening.

The above spectrogram shows a short snippet from a whale song. In spectrograms, time is on the horizontal axis, and frequency is vertical. You can reproduce this song by tracing your finger along piano keys as if drawing the lines on scrolling paper.

typically contain less predictable patterns and are sometimes produced in contexts that less obviously relate to mating.[22]

Any sound that a whale makes will generate echoes, whether it's produced within a song sequence or as a call. Without knowing what a vocalizing whale is listening for, it's difficult to know which sounds are echoically informative. How perceptually useful any given echo or set of echoes is for a singing (or nonsinging) humpback whale greatly depends on how far the echoes travel. In ocean environments, this hinges partly on the echo's strength and partly on the humpback whale's location when it vocalizes. As discussed in chapter 4, the evolutionary forces that drove humpback whales and other large cetaceans to echolocate with songs rather than with clicks may be more related to how sound waves spread underwater than to the sexual preferences of female whales.

---

22. Although as you will discover in chapter 8, humpback whales often produce both calls and portions of song when they are swimming around a potential mate.

CHAPTER 4

Where Whales Sing

GOATHERDS LIVING IN THE CANARY ISLANDS traditionally spend their days traveling through deep valleys and across rugged mountain-tops. Before the advent of cell phones, communication between goatherds was difficult. Simply yelling as loudly as possible to converse with someone located a mile or more away is neither practical nor effective. In response to this challenge, the goatherds contrived an ingenious solution: They converted their spoken language into a whistled language.

This whistled language, called *silbo*, is essentially a coded version of Spanish.[1] Using intense whistle sequences, the goatherds can construct sentences, ask and answer questions, and even tell jokes. Loud whistles are good attention-grabbers, partly because all the sound waves produced by a whistler are concentrated at a single pitch at any given time. For reasons explained more fully in chapter 5, this helps the whistle sound wave to travel farther before becoming inaudible. Experienced silbo whistlers can have conversations across three miles or more.

---

1. The native language used in the Canary Islands.

What the goatherds discovered through ingenuity or trial and error is what whales are born knowing:[2] Not all sounds are equally useful for sending signals long distances. Plus, some environments are more conducive to achieving long-range transmission than others. Environmental features strongly limit how organisms can use sound, especially when the goal is to interact over long distances.

Whistled language works in mountainous regions because there are fewer physical obstacles between whistlers. Also, the terrain itself acts as a kind of sound tunnel, guiding the flow of sound waves from the whistler to the listener. Silbo would be much less effective in a city because the many obstacles would send the sound waves off in all directions rather than concentrating their movements.

The ocean environment has many unique properties when it comes to sound transmission. You may have experienced how voices travel when your head is underwater, say, when you're in a pool or bobbing for apples. Comparing your experiences of voices heard in air versus water might lead you to conclude that water is not a great medium for vocal communication. Au contraire! Talking in water might fail for air-honking humans, but water works well for species that have mastered producing and receiving sounds while submerged. Humans' inability to hear well beneath water means you can't really trust your air-based intuitions for how whales perceive sounds. In the ocean, what *you* hear is not what *whales* get.

Oceanic Thin Disks

My analytical introduction to the world of whale research, which consisted mainly of attempts to identify programmable methods

---

2. Through millions of years of evolution.

for sorting the sounds of singing humpbacks, served as a stepping stone to my dissertation project. I'd come to Hawaiʻi to study dolphin cognition, but there was no guarantee that the dolphins would successfully learn any of the cognitive tasks I was attempting to teach them. And so, to ensure I made it to the graduation finish line, I plowed ahead with my studies of humpback whale bioacoustics, making them the focus of my dissertation research.

Naive graduate researchers sometimes do weird things, which could explain why after the dolphin sonar expert Whitlow Au summarily dismissed my hypothesis that humpbacks use songs as sonar, I asked him to serve on my doctoral dissertation committee. Members of a dissertation committee are the people who decide whether a grad student's scientific efforts are good enough to earn them the title "doctor." From my perspective, it was better to have an acoustician evaluating my research on acoustics than to have a child psychopathologist deciding my fate.[3]

Whit was, at first, reluctant to become a member of my dissertation committee, given that he thought some of my ideas were a bit bonkers. But Lou Herman and Neil Frazer[4] convinced Whit that he didn't need to agree with all my interpretations or conclusions to assess my scientific approach and the validity of my results. With those assurances, Whit signed on.

I have since supervised several doctoral students. When I advise them on selecting committee members, I never recommend the strategy of recruiting an expert who disagrees with your approach and interpretations. The principle might be sound, but the practice can get messy. Ultimately, the public defense of my dissertation lasted over three hours, in large part because Whit challenged

---

3. Another member of my dissertation committee was, in fact, a child psychopathologist. Doctoral students in psychology are typically required to include at least one psychologist on their committee who is not a specialist in the area being studied, and I chose the chair of the department.

4. Also on the committee.

many of my results. In the end he approved my project, but I don't think it was much fun for either him or me. My dad, who attended the defense, later told me, "I was this close to getting up and strangling that one guy." I guess I got what I asked for!

After leaving Hawai'i I focused on studying brains, rather than whales or sounds in oceans. Specifically, I worked on developing computer models of neural interactions between different brain regions that occur when rats, rabbits, and humans learn new associations. I was a bit surprised, then, when I received an invitation to give a scientific talk at the Acoustical Society of America[5] on humpback whale bioacoustics. The invitation was not from any whale biologist, however, but from an ocean acoustician. Since no one else was inviting me to give talks,[6] I immediately said yes. Then I started reviewing my dissertation to remind myself of what I used to know about ocean acoustics and the sounds of whales.

At the Acoustical Society meeting, I spoke about computer simulations of how humpback whale songs travel underwater, without any mention of what singers might hope to achieve by singing. This talk received a more positive reception than my first conference presentation in Hawai'i. Several attendees even encouraged me to submit my findings for publication.

Given the past failed attempts to get my whale song research published, I was skeptical that any whale-related paper I submitted would ever make it into print. Ultimately, I decided that the cost of one more rejection was relatively low, so I submitted a written version of my presentation. Coincidentally, the editor who received my submission was just starting his first year as an editor at the journal—and his name was Whitlow Au. Although this might seem

---

5. The Acoustical Society of America is a scientific society that holds conferences where all kinds of acousticians present their most recent findings.
6. Which is the norm for junior scientists.

like a worst-case scenario, it wasn't up to Whit alone to evaluate the paper's worthiness. His main role was in selecting the scientists who would judge it. Here, Whit's expertise in acoustics proved beneficial. He selected ocean acousticians to review the paper rather than marine mammal biologists. The ocean acousticians were fine with my paper. They thought my results were a bit on the boring side but considered them good enough for publication. And so my first substantive scientific contribution to the field of whale bioacoustics was born!

What did this first scientific salvo on whale song say? Basically, that sound traveling in the relatively shallow coastal environments where humpback whales often sing behaves in ways more complicated than you or anyone else can imagine. The only way scientists can get a handle on how sounds travel in these shallow-water environments is through sophisticated computer simulations.[7] Predicting how sounds will propagate—a fancy word for spread—in shallow coastal waters is a bit like predicting the path and effects of a hurricane.

Sound propagation in the ocean essentially involves a cascade of vibrations, mainly across salt and water molecules. How these molecules vibrate and how their vibrations spread to nearby molecules is not simple. Lots of factors affect what vibrates when. In shallow water, the most important factors for propagation are the boundaries surrounding the water. Those include air at the ocean's surface, rocks on the ocean floor, and nearby landmasses like shorelines. Variations in the temperature of the water, the ratio of salt to water molecules, and water depth[8] also strongly affect how sound waves propagate. This is just the beginning of a

---

7. Which are still much simpler than the real world.

8. How deep each particle is under other particles is particularly important since particles become more densely packed the deeper you go, in turn affecting how vibrations spread between the particles.

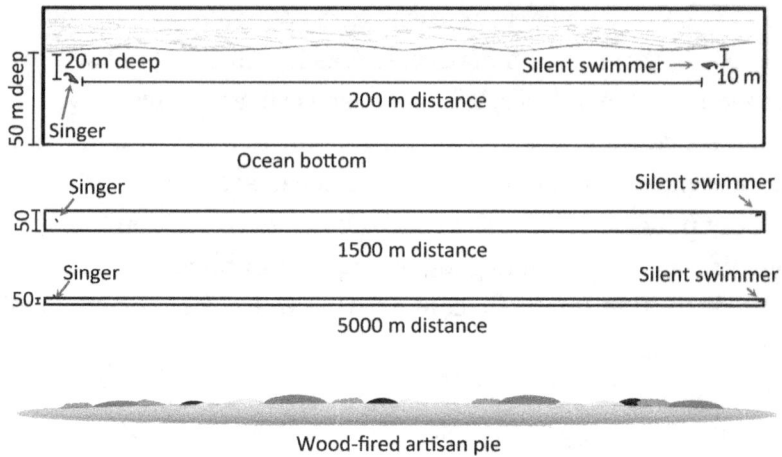

50 m deep | 20 m deep | Silent swimmer → | 10 m
Singer | 200 m distance
Ocean bottom

Singer | Silent swimmer
50 | 1500 m distance

Singer | Silent swimmer
50 | 5000 m distance

Wood-fired artisan pie

Most singers studied to date have been recorded in relatively shallow coastal waters (less than 100 meters deep), positioned 10–30 meters below the surface. Other nonsinging whales also tend to swim or rest at depths less than 30 meters. Singing in such places leads to very complex patterns of sound transmission, even at relatively short distances (short for a whale is two football fields). The farther away a silent whale is from a singer, the "thinner" the water channel through which songs travel. Sounds travel in all directions away from a singer, so you can think of songs as being broadcast from within the center of a thin pizza.

long list of features that can affect where sounds will go when produced off a coast!

To understand how songs travel in the world of whales, it's best to think of a pizza: not a pan pizza or Chicago-style deep dish but a Manhattanite, thin-crust pie. The crust of this pizza is the diameter of a small town and is made of salt water. Singers live and listen within this crust.

Now imagine you embed a tiny speaker in the center of the pizza and begin playing your favorite songs through this earbud for mice. What will happen inside the pizza? You will hear no songs coming out of the pizza, but that doesn't mean the crust isn't filled with sound. The sounds are simply trapped inside the crust. Like sounds traveling in the shallow waters surrounding coastlines, sounds trapped in pizzas do weird things.

Although the physical properties of the ocean environments noted above strongly impact how songs propagate, singing whales are not mere victims of circumstance when it comes to their songs' destinations. A singing whale has more control over how its vocalizations propagate than you might think. Most obviously, singers can control the intensity of songs. Many sounds within whale songs are loud, but some are a good bit quieter. These quieter sounds tend to fade away faster and travel shorter distances. A singer's depth below the surface also has a major effect on how songs propagate. A sound produced at a fixed loudness will travel differently depending on whether the singer made it at a depth of five versus thirty meters.

To further complicate matters, the pitch of the sound also plays a role. A low-pitched sound[9] may travel quite far if a singer is at the right depth but die out more quickly when the singer is near the surface.[10] Singers can, in principle, choose where in the world they sing, their depth when singing, and the loudness of specific sounds, as well as the pitches they produce. In this way, singers may precisely control which sounds go where underwater.

In summary, a combination of physical factors and psychological mechanisms determine how songs travel in ocean environments: the physics of propagating vibrations and whatever mental faculties whales bring to bear when selecting where to sing, when to sing, and what to sing.[11] This combination of factors is especially important to keep in mind when thinking about the sonar hypothesis because where songs travel determines not just what other listening whales might hear but also what echoes a singer might receive.

---

9. Like the low notes on the left end of a piano.

10. Sounds produced near the surface propagate in complex ways because of the air-water interface, the turbulent boundary, and variations in temperature and salinity.

11. The focus of chapter 5!

The details above relate to how sounds travel in oceans, generally. But what's the situation in places where whales actually sing? In chapter 3 we saw that whales sing in oceans around the world. When it comes to studying whale song, however, most scientists have focused on one species (humpback whales) at a few locations where they are particularly accessible: near Hawai'i, Australia, and the Caribbean. In these locales, singing humpback whales are clearly audible on hydrophones located 5 kilometers or more from the singer. In some cases they can be heard as far as 100 kilometers away. Even more impressively, singing blue and fin whales have been recorded from distances of several hundred kilometers! Given these long ranges and the fact that many singers aren't physically close to any other whales, it's safe to say that songs function over long distances.

Although most studies of singing humpback whales have been conducted near popular vacation destinations, whales don't just sing near tropical beaches. Scientists are beginning to pay attention to the songs whales produce in colder regions. So far they have found some intriguing differences. Songs produced nearer to the poles tend to be shorter and simpler than those produced in the tropics. And a few indirect observations[12] of whales singing near Antarctica and Greenland led to a surprising discovery: Some humpback whales sing while actively foraging.

Presumably, these singers are not attempting to seduce fish, so what *are* they doing? Biologists suggest that perhaps the singers are practicing in order to get a jump on the competition before their long migration to breeding areas. In other words, they think that these humpback whales are males that sing while feeding to increase their breeding opportunities.

---

12. Observations in these colder waters are mostly made using recordings collected when whales are not visible (so that what singers are doing while singing is unknown). But in a few cases, tags placed on foraging whales have recorded them singing while they continue to move in ways that indicate they are foraging at the same time.

If whales use songs for sonar, this suggests a different possibility: that feeding singers are scanning their environment for other whales. Songs used in the feeding grounds differ somewhat from those produced in the tropics, maybe because of the differences in how sounds propagate in colder locales or because of differences in how singers monitor their surroundings. In the context of feeding, knowing where other whales are located may be relevant to foraging strategies, both for avoiding collisions and for distributing effort. Such knowledge could potentially benefit both males and females. So if females forage similarly to males, then females should also be observed singing occasionally while feeding. No one has yet attempted to survey the sexes of feeding singers.[13] If female humpbacks are found to sing while feeding, then whale biologists will be forced to reconsider their assumptions about why whales sing.

Singing in the Arctic can provide important information beyond just the locations of silent whales. For example, singing bowhead whales migrating along the coast of Alaska appear to use echoes to determine the thickness of ice so they can avoid swimming underneath ice sheets that are too thick for them to break through. An inability to detect thick ice prior to swimming beneath it would put whales at risk of suffocation. It's not known from how far away singing bowhead whales can detect thick ice sheets or how precisely they can judge ice thickness. Like other solids, ice can generate strong echoes from songs, enabling singers to plan safer swimming routes. In short, songs produced in polar regions may enable singers to avoid collisions with ice and to coordinate their movements relative to those of other whales.

You may recall from chapter 1 that the word "sonar" was originally an acronym—SONAR—abbreviating the phrase "SOund Navigation And Ranging." The "ranging" part involves using echoes to

---

13. It's easier said than done, and everyone assumes the singers are males, so why bother?

reveal the distance to an object. Commercial depth finders use sound to range the water-earth boundaries below boats. When a bowhead whale uses songs to avoid thick ice, that qualifies as ranging. Navigation involves planning and following a route. Bowheads singing near thick ice appear to be navigating with songs by choosing where to swim based on the echoes they hear. They range to determine how far thick ice persists, then navigate by avoiding extensive thick sheets. In essence, singing bowheads plan where to swim to avoid future suffocation by making sure to steer clear of ice traps. Some observations of singing blue whales suggest that they can use echoes from landmasses located hundreds of miles away to aid their navigation—for instance, during long migrations.

In principle, all singing whales should be able to range and navigate using songs. Doing so doesn't require any hearing abilities beyond those that are standard issue for mammals. You could do it, too, if your ears were better suited to underwater hearing and you knew what to listen for. The question then arises, If singing whales can range and navigate using audible sounds, then why don't dolphins sing? The short answer to that question, alluded to in chapter 3, is that we don't really know that dolphins aren't "singing" to reveal environmental features. Dolphins clearly aren't singing in ways that sound like singing baleen whales. But they also aren't traveling over the same distances as those larger migrators. Possibly, the sound sequences that dolphins use to range and navigate may share properties of whale songs that have yet to be recognized.

Dolphin echolocation, though limited in its potential detection range compared to whale songs, can be used to range targets up

Opposite page: *Singing while feeding.*
The unexpected discovery that humpback whales sometimes sing while actively pursuing prey raises new questions about the nature of whale songs.

to 800 meters away. Maybe that's as far as a dolphin needs to scan to navigate effectively. Or maybe the calls and whistles used by dolphins—further described in chapter 8—provide the same kind of echoic information as whale songs, without requiring dolphins to keep repeating specific sound patterns like whales do. Bats use different kinds of sounds to echolocate depending on their needs, and the US Navy can also choose from multiple sets of sonar signals. There's every reason to expect that different species of cetaceans would also vary their sonic navigation strategies depending on their circumstances.

At this point you might be wondering why baleen whales can't just use clicks to echolocate like toothed whales do. After all, clicks seem to work fine for sperm whales, so it can't be that baleen whales are too big to click. The answer comes from the physics of sound propagation. A sperm whale's head is so bulbous for a reason—the extra forehead increases its capacity to boom out clicks. Sperm whale clicks are loud enough to travel several kilometers in the deep waters in which they vocalize. But when a click (and its echo) travels that far in relatively shallow environments—like those where humpback whales often sing—every reflection from the surface and bottom effectively produces a new echo. Clicks in shallow waters would produce trains of echoes from faraway targets overlapping in time with echoes from the ocean surface and floor.[14] Separating out which echoes were generated by an animal located kilometers away in coastal waters is probably impossible. This is less of an issue when a sperm whale is using clicks to echolocate long distances in deeper water because there will be a more direct (and usually shorter) path between the whale and potential targets of interest. Similar factors explain why most echolocating bats don't use clicks to echolocate.

---

14. Also, the higher frequencies in clicks tend to fade out faster during transmission, so echoes from distant targets are less likely to make it back to the whale.

Each species of baleen whale has a distinctive song. Some songs, like those that blue whales sing, are much simpler than those of humpback and bowhead whales. These species differences might relate to the distance that whales need songs to propagate, variations in underwater environments, and/or the reflective properties of behaviorally relevant targets. If all whales' songs were as boring as those of fin and blue whales, then it's doubtful that any scientist would have ever described whales as singing. This raises the question of why it is that some whales sing fancier songs than others. Could *where* whales sing shed some light on *what* they sing?

Ecological Origins of Song Complexity

My unexpected first publication on humpback whale bioacoustics indirectly addressed the question of how the environments within which whales sing might influence the types of sounds they produce within songs. Because underwater conditions affect how and where different songs propagate, the specific kinds of songs a whale produces within each environment determine their functions. This first paper did not, however, directly address what those functions might be. My contribution to the world of whale bioacoustics might have ended there were it not for the fact that around this same time Neil Frazer received an invitation to submit a paper describing his research on ocean acoustics.

Invited scientific papers work a bit differently from the standard research paper because a journal editor specifically selects the invitee to write on their area of expertise. Consequently, editors tend to be more flexible with what they're willing to accept from the chosen author. Neil asked me about working with him on a paper discussing the sonar hypothesis in humpback whales. I was dubious that such a paper would make it through the review process but fully on board with making the attempt.

NIGHTINWHALES?

Canaries were the first songbirds explicitly compared to singing whales, but they were far from the last. Another notable singer, the nightingale, is often described as singing in ways reminiscent of whales. For instance, in his book *Thousand Mile Song*, David Rothenberg offers, "If you speed up a humpback whale song it sounds just like a bird. It has the tonality of a catbird, perhaps, with the rhythmic precision of a nightingale."[15] Describing structural features of humpback songs, he writes: "This kind of structure for a complex song composed of repetitive parts has also been found in a few species of birds, most notably the nightingale."[16]

Nightingales produce some of the most complex songs in the animal kingdom. Their songs include more intricate patterns and sound combinations than those produced by canaries. Although nightingales' songs sound fancy, they mostly fall into four generic types that the birds have consistently produced for at least a century. Like canaries, male nightingales use their songs to attract mates, with whom they form monogamous pair bonds, and to defend territories. Although male singers get the most attention, female nightingales also sing. Females aren't singing to attract mates but may sing in the presence of a mate.

It's true that you can make a singing nightingale sound like a whale by slowing down a recording of its song. But no amount of slow-mo is going to make the behaviors of singing nightingales and singing humpback whales similar. Humpback whales and nightingales both produce complex sound sequences that one can artificially make sound similar. Of course, if you slow

---

15. See page 6 of *Thousand Mile Song*.
16. From page 351 of *Thousand Mile Song*.

down recordings of an echolocating goblin bat searching for food, it will also sound like a singing humpback whale.

Vocal sequences produced by different species often sound similar because most vertebrates vocalize using similar mechanisms—specifically, they all push air past vibrating membranes (chapter 6 provides the specifics for humpbacks). In the same way, a chimpanzee hitting a stick on a tree trunk may produce similar sounds to those produced by a human drummer, but this does not imply that the chimpanzee is hitting the tree to make music.

We spent months working out how best to summarize the evidence we'd collected, including elements of my dissertation, the remnants of previously rejected papers, and even parts of my original presentation that had provoked such hostile reactions back in Hawai'i. Neil submitted the paper to the editor and then we waited. As expected, some of the reviews were openly hostile. But other reviews, coming from ocean acousticians, were more encouraging.

In the world of scientific journal editing, this is the equivalent of a hung jury. Normally, editors would pass on publishing such a paper. But because the paper was invited, the editor decided to add a couple more reviewers to the mix. The two new reviewers were also split in their decision. This led to a bit of back-and-forth between Neil and the editor, prompting the editor to solicit two more reviews. Again, the reviewers offered opposing opinions! The paper had by now been considered by eight experts—the most I have ever witnessed for any scientific paper—with no clear consensus in sight. Ultimately, the editor decided to let us submit a revised version of the paper that addressed the critiques of the crew of reviewers who had recommended that the paper be booted. We did so, the editor was appeased, and the paper was published!

The dissenting reviewers were, unsurprisingly, not happy about this outcome. In a series of email exchanges with Whit Au, he voiced several concerns about the content of the paper. Neil and I encouraged him to publish his concerns if he felt so strongly that our paper was flawed. I especially wanted Whit's counterarguments to be published in the scientific literature. To my mind those critiques were based mainly on assumptions about what humpback whales could or couldn't do. I wanted my critics to publicly present every argument they had against the idea so that others could judge for themselves whether songs might be useful as sonar. Whit agreed and, along with several other bioacousticians, published a comment called "Against the Humpback Whale Sonar Hypothesis" to refute our article.

At the core of Whit's critique were two main ideas. First, he argued that the acoustic environments within which humpback whales sing are too noisy and complex for a singer to distinguish echoes of its own song from songs produced by other singers. Per his commentary, "How a singer could distinguish an echo from direct signals from several other singers is beyond our comprehension."[17] Second, he argued that the songs produced by humpback whales are too variable to work as sonar signals, noting that "if song were used for sonar, one would expect that the process of natural selection would result in convergence to an optimal signal."[18]

Hidden beneath the arguments of Whit and his coauthors was the counterproposal that the optimal signal for male whales to produce if they want to mate is a complex song that females find appealing. This well-established explanation for why whales sing makes female choice the driving force behind the properties of

_____

17. Au and colleagues, in their paper "Against the humpback whale sonar hypothesis," page 297.
18. "Against the humpback whale sonar hypothesis," page 299.

songs used by whales. Consequently, it says little or nothing about how environmental factors might have shaped the evolution of whale songs. A corollary of this mainstream interpretation is that different species of baleen whales sing songs differently because females' preferences are species-specific.

There's currently no way to scientifically establish what female whales like or dislike about the songs they hear, either in general or with respect to what a given singer is doing at any particular time. There will never be a way to scientifically test how whale songs evolved their properties. But it is possible to measure how songs propagate in the environments where whales sing. And, as a result, scientists can determine whether the sounds within whale songs are "suboptimal" for generating informative echoes.

First, let's consider the issue of noisiness. Oceans can be noisy places. On the other hand, singers are quite loud. It's not at all hard to hear them when they are booming. The sound sources likely to cause problems for a singer listening for echoes are other vocalizing whales and humans piloting large ships. Evolutionary processes had no opportunity to shape whale hearing or behavior to accommodate the intense noises that humans constantly spew into the ocean.[19] In contrast, those processes did have millions of years to shape humpback whales' ability to deal with sounds made by other whales.

Levels of whale-generated noise vary greatly depending on the location, time of day, and season. Some of this background hum comes directly from the singers themselves, but much of it is the result of persisting environmental echoes, called *reverberation*. If you've ever been in a large gym containing a basketball court or swimming pool, you've experienced reverberation. Based on those experiences, you can probably imagine how reverberation might interfere with hearing and interpreting sounds.

---

19. A central topic of chapter 10.

Ocean acousticians and bioacousticians alike have treated reverberation from whale songs as a natural source of ambient noise
that marine animals must overcome when using vocalizations to
communicate. You might be surprised then to learn that singing
humpback whales appear to consistently produce sounds within
songs that are exceptionally prone to reverberating. In fact, some
sounds that singing humpbacks produce may reverberate for
fifteen seconds or more within the environments where they
sing. Singing humpback whales also often position themselves at
locations where reverberation is more likely to occur.[20]

Why would natural selection lead to situations where song-
generated noise levels are high if such noise is a hindrance that
whales must overcome? Why aren't singers that produce less reverberant songs in less reverberant environments more likely to
reproduce than singers who are effectively boosting their noise-
generating potential? Chapter 7 provides a possible solution to this
conundrum. For now, it's enough for you to know that singers do
not behave as if they are attempting to minimize their noisiness.

Animals have adopted a wide range of strategies for dealing with
the difficulties of using sound in noisy conditions. While human
engineers have trouble pulling signals out of noise, bats seem to be
able to overcome all kinds of noisy conditions when echolocating.
Swarms of bats often hunt insects in large groups without showing any signs of being confused about which echoes are coming
from whom or where. Experiments performed with bats in labo-
ratories similarly show that echolocating bats can handle most
natural and artificial noise sources that scientists throw at them.

This is not too surprising given the wide variety of animal-
generated sounds that exist in the natural world. In fact, species
seem to have coevolved so they not only use vocalizations that avoid
interference within their own species but also choose sounds that

---

20. Such as near the edges of seamounts.

are not susceptible to interference from other species. Scientists hypothesize that animals do this by focusing their vocal efforts on sounds that do not overlap in time or pitch with those of their neighbors, a proposal called the *acoustic niche hypothesis*. Essentially, the idea is that natural selection favors individuals that can make use of a relatively uncluttered sound channel within a shared environment. When multiple species evolve in a shared space, this shapes the vocal repertoires used by them all.

Animals that depend on their listening skills to thrive have discovered ways to overcome all kinds of environmental constraints on localizing and identifying sounds. The notion that humpback whales have not made any evolutionary strides in this regard is laughable. The issue of exactly how singers perceptually handle the complex soundscapes they create will be addressed in chapter 7. There, I will argue that the complexity of humpback whale songs is a direct result of how coastal environments affect sound propagation.

Above, I noted that where a singing whale produces sounds will have large effects on how far those sounds propagate. Many, but not all, sounds produced by singing whales can travel impressive distances. Long-range transmission in the ocean is, however, limited by more than just how far sounds propagate. Sounds traveling underwater are not like marbles rolling through a marble run. As sounds travel underwater, they become increasingly distorted. Like an ice cream sundae on a warm day, the sounds that reach a listener become more and more dissimilar to their pristine selves the longer and farther they travel. This means that even if a listener can hear a song produced by a singer 10 kilometers away, there's a good chance that components of that song will be either garbled, uninterpretable, or entirely missing.

Propagation-related song distortion poses a problem for any listening whales attempting to assess the fitness of singers based on their songs because the listener cannot determine what the original songs were like before they became distorted. It would be akin

Whale Song Acoustics: Songs as Sirens

One way that humpback whales use song is to home in on singers. The main, and possibly only, whales that make a beeline (whaleline?) toward singers are males. Usually, the joiners only hover around the singer for a few minutes before heading off. So it's not clear why they join the singer. Often, the singer stops singing during the visit. Other times the singer just keeps plugging away. Occasionally, the singer will head out with the joiner to seek new adventures. Maybe males that approach singers are looking for potential partners in crime?

Whatever the reasons, these incidents make it clear that songs provide homing signals to other whales. As a homing beacon, most of the fancy features of whale songs are unnecessary. If all a singer wanted was to announce their location, they could do so easily by honking loudly every few minutes. And if singers really wanted to maximize the range at which other whales could locate them, then collecting more honkers in one place would extend their reach. Having more singers in one locale might then draw the attention of other whales, like the sounds of an orchestra warming up or the potpourri of music coming from a carnival.

Lou Herman argued that this sort of "acoustic invitation" might allow humpback whales to gather in different places in different years; wherever the most singing is heard, other migrating whales should head. Similar arguments can be made about whales singing while migrating. Whales could get info about changes in migratory paths from listening to singers at the vanguard. Finally, some have suggested that the spatial information provided within songs may enable singers to stake out a claim around the spot where they are singing, establishing a temporary territory. Singers do seem to space themselves apart when in breeding areas. So songs may both attract silent males and repel singing males. Usually, a singer that approaches another singer stops singing before it gets too close.

Describing songs in terms of their spatial origins is much simpler than describing their structure or musicality, in principle. For a whale trying to home in on a singer or singers, though, the relevant information is more about where the positions of the singers are relative to the listener's current location rather than about where they are in any geographical coordinates. Also, for a human listening to a singer underwater, figuring out where the song is coming from is much easier said than done! Thinking about how songs reveal singers' locations is probably closer to the heart of understanding what songs are like for whales than appreciating their aesthetically appealing qualities.

Distant listeners must use acoustic cues to figure out a singer's location.

Song traveling directly from a singer to a listener provides few spatial cues.

Sounds propagating over more complex paths provide more clues to a singer's position.

The ability to localize sounds is a fundamental feature of mammalian hearing. In the case of humpback whales, song may serve as a beacon that can attract others like a dinner bell or repel them like a lighthouse repels ship captains. To function in this way, songs merely need to be conspicuous.

to trying to figure out the original shape and artistic quality of a candle after it has melted. This is part of the reason why many whale song experts emphasize the patterns of sounds that singing whales produce rather than the properties of the sounds within those patterns. If the details of sounds within songs are arbitrary and irrelevant to listeners, then it doesn't matter so much if some of those details become distorted. Importantly, the kinds of distortions that create challenges for distant listeners have the opposite effect for singers attempting to interpret returning echoes from songs. This is because the singer senses every sound it produces prior to any propagation-related distortion. Specific neurons in the brain of a singer may, in principle, compare the distortions within received echoes to the properties of the original signal. Consequently, singers can potentially extract and identify all sorts of details about the distance and direction of the sources of those echoes. This is what echolocating bats appear to do as they process incoming echoes generated by their ultrasonic cries.[21]

The distorting effects of coastal ocean environments set up a kind of *experimentum crucis* for the functions of whale songs. An *experimentum crucis* is a scientific scenario in which different hypotheses predict opposite outcomes, thus providing a way to construct a "death-match" between rival interpretations. A famous historical example of this relates to tests of the wave theory of light. The once dominant particle theory of light predicted that if you shone a light through two slits, then only the particles of light that made it through the slits should reach the other side, and you should see two lines of light (like spray painting through a stencil). The competing wave theory of light, however, predicted that if light consisted of waves, then the two slits should create two new sources of waves that would interfere with each other, leading to

---

21. Chapter 5 reveals a suite of surprising similarities between the sounds that bats use to echolocate and the sounds of singing whales.

multiple light and dark bands on the other side of the slits (an interference pattern like those created by transmitting sound underwater, but simpler). The results of the double-slit experiment were conclusive—anyone could see that there were more than two bands of light on the other side of the slits. The particle theory of light was vanquished![22]

The *experimentum crucis* for whale song is this: If singers are striving to create displays that enable other listening whales to assess their fitness, then they should produce songs that retain as many features as possible after traveling long distances so that listeners can clearly receive their message. After all, the differences in songs produced by two closely matched competitors are likely to be relatively subtle, such that any distortions might tip the scales in terms of which singer is judged the fittest. If, however, singers are producing songs to provide themselves with information from returning echoes, then it would be to their advantage to produce songs that are prone to the distorting effects of long-range propagation because this can increase the amount of information available from each echo. The two hypotheses—song as display and song as sonar—predict incompatible outcomes such that only one can emerge victorious once the properties of songs are revealed.

It would be nice if the studies necessary to reveal the winner of this scientific competition were as simple as shining a flashlight at a piece of metal with two slits in it, but alas that is not the case. The waves that singing whales emit are much more varied than those in a beam of light, and the environments through which songs travel equate to a million overlapping slits. Also pertinent is the slight complication that humpback whales continuously vary the sounds and songs they produce throughout their adult lives. As a result, the outcome of the *experimentum crucis* in one year

---

22. Not really. In the world of physics and science more generally, it's rare that one experiment can settle a dispute. The *experimentum crucis* is more of a holy grail.

might, in principle, differ from the outcome in the following year (a wrinkle that will be considered more closely in chapter 8). Despite these snags, it's clear that understanding how singing whales vary the acoustic characteristics of the sounds within songs is fundamental to working out why they are singing.

In chapter 5, I will do my best to reveal the true physical nature of humpback whale songs, showing not only how the structural and acoustic features of songs can enable singers to distinguish self-generated echoes from the direct signals of other singers but also how song variations can optimize the utility of returning echoes, as one would expect based on processes of natural selection.

CHAPTER 5

What Whales Sing

Her high sharp cries
Like shining needlepoints of sound
Go out into the night and
echoing back,
Tell her what they have touched.
She hears how far it is,
how big it is,
which way it's going:
She lives by hearing.

—Randall Jarrell, A Bat Is Born

THE EARLIEST RECORDINGS OF SINGING humpback whales made public by Roger Payne have a haunting, otherworldly quality that provokes a sense of wonder in many who hear them. The intricate structures, rhythms, and melodies of whale song are surprisingly sophisticated. Listening to just a single song reveals the aesthetically appealing and impressively complex qualities for which these unique vocalizations are famous. The extreme diversity of humpback whale songs, however, is what has attracted the most scientific attention, demonstrating a variety that can only be appreciated by observing songs collected from many different times and places.

In listening to humpback whale songs recorded from multiple populations across five decades, I've gradually reached the conclusion that this approach is a lost cause, scientifically. Listening to whale songs is fine for personal inspiration or perhaps meditation. But simply hearing different songs is not that helpful for figuring out songs' functions. I noted in chapter 1 that one of the first and most important clues that led me to suspect singing whales weren't vocally strutting was the fact that, unlike any songbirds, singing humpbacks continuously change their vocal repertoires as adults. This phenomenon, combined with the complexities of sound propagation described in the last chapter, should be enough to convince scientists that singing whales aren't just conversing or acoustically flexing. In most other mammals that use acoustic displays as a mating strategy, evolution has favored vocal stability. Even dolphins, who vocally interact over much shorter distances than singing humpback whales,[1] rarely change the sounds they use as adults.

I've concluded that listening to humpback whale songs is the wrong way to go because what humans experience when hearing those songs is not what the whales themselves hear. According to the sonar hypothesis, what singing whales perceive differs so radically from our perceptions that we might as well be sensing totally different events. That's because what's important to a whale using songs as sonar are the silent intervals following the sounds within songs. Humans perceive these gaps as "silence"—any differences between consecutive silences are thus imperceptible to listening humans. But from the perspective of the sonar hypothesis, what

---

1. Meaning dolphins' vocalizations are less likely to suffer as much propagation-related distortion as the sounds within humpback whale songs.

humans perceive as being nothing means everything to singing whales.

British subway stations are known for their cautionary slogan "Mind the gap," meaning "Avoid letting your foot enter the space between a train and the platform, because if you do you will likely regret it." In the case of whale song analyses, a better slogan would be "Mind the gaps." In other words, "Pay attention to those nothings between the sounds, because if you don't, you will likely remain confused about why whales sing."

Variations across the "silent" gaps between sounds are critical for any animal attempting to both detect and localize echoes from moving targets. That's because the gaps are not silent at all but filled with potentially informative echoes. Echolocating animals listen carefully to returning echoes that arrive during these pseudo-silences. The echoes filling the gaps reveal the differences between nothing-there and something-there, as well as the differences between something-stationary versus something-moving. The relative changes in echoes over time are more perceptually relevant than the properties of any single echo. This is because differences in the intensity and timing of successive echoes can reveal the shapes, speeds, and evasive maneuvers of living targets. In this context it's important to closely consider how the structural properties of whale songs may contribute to the information that singers can extract from incoming echoes. Toward that goal, let me reintroduce you to an exemplary producer of perceptually useful echoes: the little goblin bat.

You may recall from chapter 3 that the little goblin bat is a Cuban resident that sometimes rhythmically alternates between two differently pitched ultrasonic sounds while searching with echolocation. If you could hear the ultrasonic cries of a little goblin bat, they might sound something like an odd cuckoo clock chiming—the higher-pitched cry being the "cuh" and the lower-pitched one being the "coo." When an echolocating goblin bat

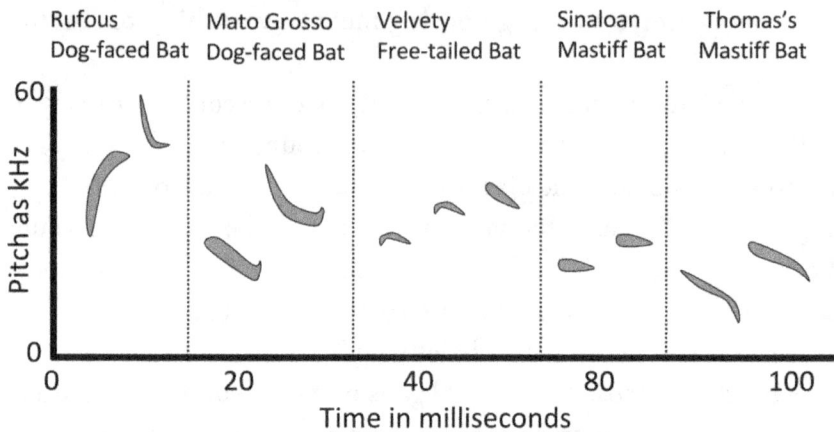

| Rufous Dog-faced Bat | Mato Grosso Dog-faced Bat | Velvety Free-tailed Bat | Sinaloan Mastiff Bat | Thomas's Mastiff Bat |

Several species of bats produce alternating echolocation signals, some-times called plip-plops. Plip-plopping bats rhythmically alternate be-tween two different echolocation calls while searching for targets at long distances. Bats may alternate the intensities, durations, spacing be-tween calls, pitches, and/or forms of their calls while searching. The spectrograms above provide a sample of the kinds of sounds these bats alternate between—different species of bats alternate between cries with different acoustic features. Some bats, like the velvety free-tailed bat, even cycle through three cries (plip-plop-ploopers?).

alternates pitches like this, the echoes from those sounds will re-flect those changes, creating two sets of echoes: a stream of high-pitched echoes and a stream of lower-pitched echoes. Several other species of bat also alternate echolocation signals in this way, including the eastern red bat, the dog-faced bat, the mastiff bat, the free-tailed bat, and the hoary bat. Some bat researchers call these kinds of alternating search signals *plip-plop calls*.

Scientists have studied echolocating bats more intensively and for a lot longer than singing whales. Yet they still aren't sure what bats achieve by plip-plopping. Most assume that the bats are either alternating call features to segregate the streams of resulting echoes—making it easier to match emitted sounds to returning echoes—or that the different calls provide different kinds of echoic information.[2] Most general descriptions of bat sonar make no mention of bats searching with plip-plop calls. I'd certainly never heard of them when I started grad school. I only discovered plip-ploppers because I was looking for them. And I was only looking for them because I was trying to understand why singing humpback whales would produce such a wide variety of sound patterns if they were simply trying to generate echoes from other whales.

My initial intuition was that if patterned sonar signals work well for singing humpback whales, then patterned calls should potentially be useful for other echolocating animals. I was stunned, however, when I first came across plip-plopping bats. These bats not only produced sonar signals in patterned sequences while searching for targets at long distances but had call patterns that looked spookily similar to those present in humpback whale songs.[3] Discovering this unsuspected convergence in vocal behavior across echolocating bats and singing whales further convinced me that humpback whales were singing in ways that could generate informative echoes and that their songs might be more sophisticated echo generators than the simpler patterns produced by bats.

The most important consideration when analyzing the echo-generating potential of whale songs is where to focus: Will you attend to the audible features of songs or to the not-so-silent gaps between sounds? Traditionally, researchers have emphasized the structural complexity of songs and how that structure varies over months and years. This strategy has left researchers blind to the precise ways in which singing whales constantly adjust both gaps and streams of sounds within their songs. By controlling the successive echoes generated by their songs, singing humpback whales

---

2. For instance, some echoes might reveal the positions of other bats or a bat's distance from the ground, while others might be specialized for detecting distant insect swarms.

3. I will reveal just how similar these sound sequences are in chapter 6.

may be able to systematically scan large regions of their surrounding environment. Like a movie director using different angles and camera shots to create dramatic scenes by manipulating the visual elements that the audience sees, singing whales may construct their own "echoic scenes" using precisely organized sound patterns, thereby revealing social dramas unfolding at the edges of their perceptual horizons.

A Morphing Mix of Rhythmic Tweets and Moos

Although humpback whale songs are structured in time, many of the rhythms that singers produce are too slow for humans to easily perceive. The slowest tempos that humans can recognize as having a beat are between 30 to 40 beats per minute. The typical tempo of a humpback whale song is around 13 beats per minute. You'd be hard-pressed to clap steadily at that rate for even 1 minute. Singing humpback whales, however, do it daily for hours on end, like giant oceanic metronomes.

The temporal steadiness of song production is one reason why other scientists are skeptical that singers are echolocating. When dolphins (and bats) echolocate, they gradually adjust the durations of silences between sounds, leaving enough time for echoes from one sound to return before producing the next sound. Usually. But recall from chapter 1 that when dolphins scan objects at long ranges, they switch to using packets of clicks, which they produce at a more rapid, rhythmic pace so that echoes from one click may overlap with echoes from prior clicks. Plip-plopping bats also produce echolocation cries at a steady pace when they are presumed to be searching for faraway targets.[4]

---

4. If at this point you are thinking that bats are too different from baleen whales for any meaningful comparisons between these groups, you might be surprised to learn that, aside from the ungulates, bats are the closest genetic relatives to whales.

Side view

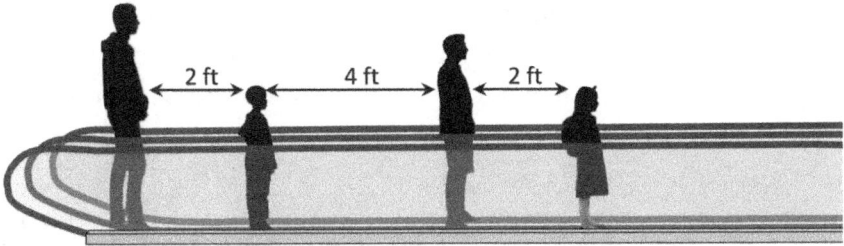

Top view

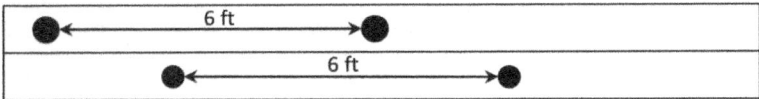

The alternating cries produced by plip-plopping bats (see spectrograms on page 96) are like riders on two adjacent moving sidewalks. Although the spacing between riders appears to alternate when viewed from the side, viewing the riders from above reveals the spacing on each sidewalk to be the same. Plip-plopping bats and singing humpback whales produce regularly timed sound patterns that appear rhythmically structured. Those rhythmic patterns can alternatively be viewed as two simpler streams of sounds that overlap.

Visualizing how plip-plopping bats time their alternating cries may help you better understand the structural features of humpback whale songs. First, imagine two moving sidewalks beside each other going in the same direction at the same speed. Now imagine that each sidewalk has an attendant controlling when people get on. Let's say these moving sidewalks are provided for fathers and their progeny, with one sidewalk exclusively for the fathers and the other exclusively for their children. In this visualization, the children are the searching bats' "plips," and the dads are the "pops."

The attendants' jobs are to make sure that the dads step onto their own sidewalk such that they are exactly two feet behind their child and six feet behind the father in front of them. To achieve this goal, both attendants precisely time when the children and their fathers step onto the sidewalks. A camera placed overhead would reveal that individuals on each sidewalk are regularly spaced. The physical spaces between riders correspond to the durations between plips and plops—the times when arriving echoes can potentially be detected by bats.

Now imagine that various groups use the sidewalks at specific times of day, always arriving in the same order. For example, kids and their dads use the sidewalks from 8:00–8:30 a.m., moms with teens from 8:30–8:45 a.m., moms with elderly parents from 8:45–9:30 a.m., teen clans between 9:30–10:00 a.m., and so on. If the attendants give different boarding instructions based on which passengers are present, then peoples' spacing on the two sidewalks will vary for different groups. This expanded example takes us beyond the aerial scans of plip-plopping bats to the more variable vocal sequences produced by singing humpback whales.

The humpback whale songs that Roger Payne and Scott McVay originally described in the early 1970s show patterns analogous to those of the parading parents described above. First come motorboat sounds, each followed by a tone or two. Next in the parade are high-pitched chirps. Following a slew of these come pairs of loud, rising whistles, accompanied by faint high squeals. Then there are five roars and a grunt. The last two groups consist of sustained tones, falling warbles, grunts, and the occasional "pulsive note." Whenever a singer makes it through this last group of warbles, it restarts from the beginning with more motorboats and tones. This ordered sequence of grouped sounds is what the Paynes christened a "song." If you're having difficulty forming a mental image of this convoluted acoustic cavalcade, then you can appreciate why it took scientists so long to discover the patterns within humpback whales' vocal sequences!

Although Scott and Roger noted that many of the sequences produced by singing humpbacks were "monotonously repetitious in rhythm and frequency,"[5] they made no attempt to analyze the timing, tempo, or cadence of sounds within songs. Instead, they focused on the sets of repeating sound patterns within songs, which they referred to as *themes*. Using that terminology, the song described in the paragraph above contains six themes produced in a standard order, with Theme 1 being repetitions of the motorboat with tones pattern.

Roger and Scott defined each theme based on the repeated sound patterns within it. For example, if Theme 1 contained 15 "dads" with their sonic kids, it could be split into 15 parts, referred to as *phrases*. Each phrase was then further dismembered into its constituent sounds, which they referred to as *units*. The number of times a singer repeats patterns within each theme varies from one song to the next. In the songs that Roger and Scott described in their initial report, Theme 1 contains between 9 and 41 repeated phrases, with each phrase lasting about 14 seconds. Different singers may spend more or less time producing each theme. Some phrases are consistently repeated less often than others, however. For instance, Theme 4, as described in the inaugural report on whale songs, always includes only a single phrase.[6]

In summary, Scott and Roger proposed an ingenious scheme for describing the cyclically repeating sound patterns that singing humpback whales produce. Over time, their descriptive framework came to be viewed not only as a way of subdividing sections of whale songs for analysis but also as a scientific claim about the structural properties of songs. Nowadays, it's common for whale researchers to state in papers and presentations that humpback whale songs are hierarchically structured. The term "song" itself

---

5. Roger Payne and Scott McVay, as noted in their 1971 paper, "Songs of humpback whales," page 591.

6. Meaning that singers consistently spent much less time producing Theme 4 compared to Theme 1.

Whale Song Acoustics: Songs as Streams

Most scientists analyze whale songs by first categorizing each sound a whale produces into a type (say a "moan" or a "snort"). Then they simply transcribe the sound types in the order they were produced by the singer. If you do this for the beginning of the song "Row, Row, Row Your Boat," you are likely to get row-row-row-your-boat. Or you could categorize the sounds differently, perhaps verb-verb-verb-adjective-noun or splashy-splashy-splashy-huggy-floaty.

This approach leaves a lot behind when applied to whale songs. You throw out the rhythm of the song, the pace of sound production, and the properties of the sounds (so that piccolos and tubas become equivalent). You also discard the relationships between the sounds within the sequence. Ask a friend to sing the phrase "verb-verb-verb-adjective-noun" and you'll see what I mean. Whatever they produce, it's unlikely to resemble the melody of "Row, Row, Row Your Boat." That's a lot of vocal detail to throw away for a species that's been honing their hearing and singing abilities for millions of years!

When you look at the actual acoustic variations within songs, rather than just the patterns of units, you start to see that singers are not simply repeating phrases. Singers are persistently changing the patterns they produce. Whales gradually adjust sounds as they progress through a song, stretching them in time, shifting pitches, merging some, and splitting others, like a sonic lava lamp.

Most scientists ignore these adjustments, preferring instead to focus on categorizing and comparing the patterns. That provides some information, but it might also miss the vocal variations at the heart of whale songs. The patterns you hear when listening to a whale song are what happens when a singer produces multiple streams of sounds in parallel. Those patterns may differ considerably from those that the singer experiences as it gradually morphs multiple sound streams.

Some singers loop songs non-stop for over twenty hours!

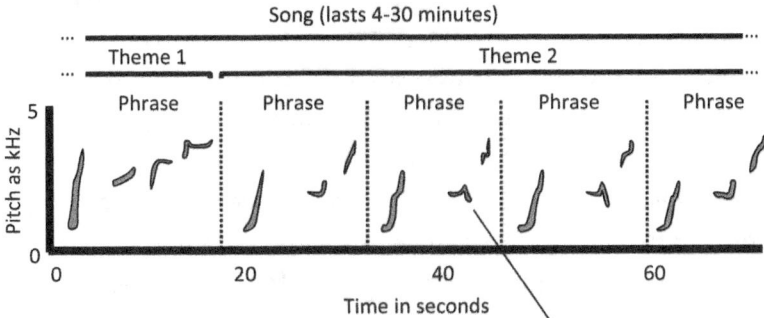

Song (lasts 4-30 minutes)

Theme 1

Theme 2

Phrase | Phrase | Phrase | Phrase | Phrase

Pitch as kHz

5

0

0   20   40   60

Time in seconds

Singers repeat phrases within themes while subtly shifting unit forms.

The spectrogram above reveals just how precisely singers rhythmically repeat unit patterns. The regular intervals between units lead to predictable visual gaps of "no-sound." Although most researchers describe whale songs as hierarchically organized, an alternative way of conceptualizing and describing the three-unit phrases shown above is to think of them as three separate streams of units (low-, medium-, and high-pitched units), with each stream being produced at a regular rate—like a dad with two kids spread out across three parallel moving sidewalks.

describes a level within this proposed hierarchy. At the top of the hierarchy is the *song session*, which includes all the vocalizations a whale produces from the time it starts singing to the time it stops. Song sessions can last many hours. Songs are simply repeated cycles of vocal patterns within these sessions. Humpback whale songs range from a few minutes long to about 30 minutes long. Songs can be broken down into an ordered sequence of themes that consist of repeated phrases containing individual units produced in rhythmic patterns.

Whale researchers enthusiastically adopted this divide-and-conquer approach to describing whale songs. Today its use is near universal. It's an intuitive approach for human observers given that we often describe speech in terms of sentences, words, and letters. It seems natural to break down sound sequences in this way, focusing on smaller and smaller chunks. The fact that researchers can describe whale songs using hierarchical terminology, however, provides no evidence that whale songs are themselves hierarchically structured.

It's true that the descriptors song, theme, phrase, and unit are hierarchical in nature: Units are parts of phrases that are parts of themes that are parts of songs, which are parts of song sessions.[7] The relevance of these demarcations to singing whales remains to be seen. No one would claim that your ability to describe your life in terms of years, months, weeks, days, hours, and seconds is evidence that time (or your life) is hierarchically structured. It's a bit sobering to realize that what whale researchers call songs might be phenomena that no whale has ever perceived. It's entirely possible that songs, themes, phrases, and units are all artifacts of human analyses.

No one knows for sure how singing whales perceive the structure of songs.[8] To keep things simple, for the moment I'll focus on

---

7. That are parts of days that are parts of weeks that are parts of seasons, etc.

8. Chapters 6 and 7 consider some likely scenarios based on what scientists do know.

units because each unit generates its own set of echoes. Applying whale terminology to the vocal actions of bats (a bioacoustics no-no), one plip is a unit, and one plop is another unit. Assuming all plips are similar, we could describe them as a unit type. It's important to keep in mind that although labeling sounds is convenient for sorting them, it can also obscure relationships between the sounds. For example, one bat species' plip might be another's plop, meaning that while acoustically the calls are the same, the plip is the highest-pitched cry for one species and the lower pitch cry in the plip-plop pattern for another species. Also, what starts off as a plip may over time transform into a plop, leaving no clear dividing line between what counts as a plip or a plop.

Unlike the plips and plops used by echolocating bats, the units in humpback whale songs are audible. Units vary in pitch, form, duration, intensity, and complexity. The units that singers produce in any given year typically can be sorted into a relatively small number of unit types,[9] such as "grunts," "chirps," or "warbles." The categorical boundaries separating each unit type are fuzzy, however, and vary from year to year.

Singers rarely repeat units across phrases without varying some of their features. More often, singers shift units along one or more acoustic dimensions across repetitions, a phenomenon called *morphing*. When singers morph units, they might stretch them out or compress them in time. They may shift a unit's pitch, gradually producing the unit at lower and lower frequencies or rapidly increasing its pitch as if they had inhaled helium. These variations in units could be involuntary, but it's not likely. Evidence that singers produce consistent trajectories of unit change across songs[10]

---

9. Usually fewer than 20.

10. The most prevalent morphing patterns involve shifting from higher-pitched to lower-pitched units and from shorter-duration to longer-duration units. No one knows exactly why singers do this; some possibilities are presented in chapter 6.

suggests that singers are precisely controlling how units morph as opposed to just being sloppy.

Because singing humpback whales morph units across repetitions, properties of phrases also change each time a singer repeats them. Not all units change at the same rate within a song, though. Some unit patterns remain relatively uniform over time, while others morph more rapidly. Singers also gradually segue from one theme to the next, linking them together with *transitional phrases* that contain elements of both the preceding and following theme. In this respect, a singing humpback is like a club DJ, gradually adjusting the timing and pitches of themes to smooth the transitions between them.

Singing humpback whales are like DJs in another way: Their sessions of continuous sound production can go on and on. Several whales have been recorded singing nonstop for over 20 hours straight. A 15-minute song can easily contain 150 units, such that a single song session might contain over 12,000 units. Unlike a rave, however, whales' song sessions contain those "silent" gaps we need to mind. How much of a 20-hour song session is sound versus gaps? On average, the time a singer spends making each unit is about equal to the duration of the gaps between units. In other words, singing humpbacks spend half their time producing units and half their time waiting to produce the next unit. This means a 20-hour session includes about 10 hours of gaps.

This calculation likely underestimates gap duration, however. Think back to the parallel moving sidewalks mentioned earlier. Viewed from the side, the observed spacing between people would be two feet between children and their fathers and four feet between those same children and the stranger-fathers in front of them, giving an average spatial gap of three feet. Viewing the scene from above, though, reveals six feet of space between each father and six feet of space between each child; the average spacing between people is thus twice what it seems like when viewed from the side. If plip-plopping bats and singing whales produce and listen to sounds within two separate streams, then the effective gaps between plops will be longer than the silent intervals between individual sounds. In other words, if singers segregate differently pitched units into two or more separate streams, then they may be able to process returning echoes from one stream independently of units and echoes from other streams.

A singing whale that is constantly morphing the properties of its units or the intervals between units might always follow the same trajectory of change. Alternatively, a singer might adjust how it modifies units in real time based on the circumstances. Overall, singers appear to morph units along predictable trajectories, with similar kinds of changes evident for different whales in isolated populations over multiple years. You could say that singers appear to obey natural laws of song-changing etiquette. For whales, however, the "proper" way to sing is more likely related to how songs function than to rules of custom.[11]

Sometimes, though, singing humpback whales morph units in ways that "break the rules." For example, a singer might backtrack and repeat a pattern that it already finished earlier in a song. Singers also sometimes flexibly shift the pitches of units up or down within a song. These changes seem to be introduced on the fly, providing evidence that singing whales not only have voluntary control over multiple song features but also constantly monitor units and tweak them in real time. In short, individual singers have a lot of control over how they gradually progress through each song.

Describing variations in sound patterns produced by singing whales is the bread and butter of whale song research. My early

11. Many researchers studying singing in humpback whales believe they are constantly copying whatever new song twists they hear from their peers, like teenagers picking up the latest slang from influencers. Chapter 8 will consider this belief more closely.

studies of humpback whale songs in graduate school could be summarized as "101 ways to measure and describe a whale song recording." Basically, I spent most of my time trying to figure out what elements of songs to measure, how to use those measures to sort different sounds and sound patterns, and how to justify using certain measures and sorting strategies rather than others. I eventually managed to publish a few of my findings on song properties about a decade after I completed the analyses, but it was a painful process.[12]

A newly hired professor at a research-intensive university cannot spend time on projects that don't lead to grants and papers without risking becoming a newly unhired professor. Consequently, I was strongly encouraged not to waste any time analyzing whale songs. But a graduate student in my department discovered that I had studied humpback whale songs in a past life and asked me to help him analyze some songs to fulfill a requirement for his doctorate. Supervising such projects is part of the job description for professors. And so I ended up supervising yet more analyses of a slew of song recordings.

Every time I've analyzed a batch of song recordings, I've seen things I did not expect to see. Each time is more surprising than the last, in part because I keep thinking I've analyzed way more whale songs than any human should and have probably seen everything there is to see in song recordings. This latest reanalysis revealed that although the audible features of Hawaiian humpback whale songs changed progressively over a decade, patterns in subsets of those features did not. Specifically, when the differences between consecutive units were analyzed rather than the acoustic features

---

12. In one case I got so frustrated with the unending diatribes of dismissive experts that I actually wrote a letter to an editor rejecting her rejection of my paper based on the nonscientific critiques provided by the most negative reviewer—amazingly enough, this strategy worked!

of the units themselves, those differences turned out to be quite stable.[13] It's analogous to a pair of siblings growing up. Lots of things about each sibling might change systematically over the years—their heights, hairstyles, clothes preferences, et cetera—but their age difference remains the same.

Why would singers continually change the properties of units and patterns of units over time but preserve the acoustic relationships between consecutive units? Perhaps something about the physiological mechanisms of song production forces singers to change properties of units within phrases all in the same direction.[14] Another possibility is that preserving the relationships between consecutive units provides some functional advantages. In other words, singers may structure phrases in ways that make the phrases work better, independently of the specific properties of the units within the phrases. This idea is similar to the mainstream hypothesis that singers change their songs to maximize their sexual prospects—it doesn't matter how you make the song fancy, as long as it stays fancier than competitors' songs. The sexual display hypothesis doesn't predict anything about what stable patterns between units should be like, however. The sonar hypothesis does because if whales use consecutive units as sonar signals, then echoes from those units can potentially overlap in time and/or pitch, making those echoes more difficult for singers to interpret. Any time echoes from consecutive sounds overlap, it introduces ambiguities about which sound generated which echoes. By making consecutive sounds within phrases distinctive, singers can potentially avoid such complications.

---

13. For example, a singer might produce a one-second unit followed by a two-second unit in one year but then produce a two-second unit followed by a four-second unit the next year. The units in the second year would sound different, but the ratio of the unit durations relative to each other would remain the same because the singer expanded the durations of both units in parallel. The temporal relationship between the units is stable.

14. A possibility I will mostly rule out in chapter 6.

Acoustic relationships between units come in two main types: pitch based and time based. The pitches of consecutive units are usually either closely matched or distinctly different. Singers also maintain a relatively fixed pace of unit production, as well as a stable phrase duration, making songs highly rhythmic. These properties account for much of the stability in song structure across years. When a singer adjusts its pace of song production, it's akin to increasing or decreasing the tempo of music—all the rhythmic relationships within the song are maintained. If a singer repeats consecutive units, then the differences between them are stably minimal.[15] And if a singer alternates between different unit types, which is something singers commonly do . . . then what? Well, then things get interesting because the specific acoustic relationships between units within a phrase will determine the echoic scenes that the singer can potentially perceive. Like some echolocating bats, singing whales may produce alternating plips and plops to segregate echoes into streams, thereby avoiding self-interference and echoic ambiguities.

To summarize, when thinking about what humpback whales sing and how this relates to why whales sing, the key factors to keep in mind are (1) the gaps; (2) that "songs" are a parade of themes, with each theme corresponding to a specific pattern of units (a phrase) that is usually repeated multiple times; and (3) themes typically occur in a looped order, with the singer flexibly morphing units and phrases in real time within each song. These properties of songs are not hypothetical. They are present in most recorded humpback whale songs. Occasionally, singing whales will deviate from these norms, and undoubtedly, many features of these songs have yet to be recorded or analyzed. Additionally, so far in this chapter I've focused on the most obvious features of songs that any observer could detect by listening to almost any

---

15. Like identical twins.

HORNY SHEEP

At the core of mainstream explanations for why whales sing is the idea that songs are sexually selected displays: Males possess an elaborate trait (songs) that females do not, and therefore the purpose of that trait is to impress other whales, thereby increasing mating opportunities. On the surface, this explanation sounds reasonable. But let's apply it to our terrestrial cetacean surrogate, the mountain sheep. Specifically, bighorn sheep.

As their name implies, bighorn sheep are adorned with massive horns. Well, not all of them. It's only the adult rams who sport the curly monsters. The ewes also have horns, but their horns are much smaller. A ram's horns grow every year so that the oldest rams usually have the biggest horns.

The notable difference in horn size between rams and ewes immediately suggests the possibility that the mega-horns are a result of sexual selection. In other words, horns may function as a visual sexual display like a peacock's tail. If larger horn size is correlated with fitness, the display could warn off rams with smaller horns and impress receptive females. Behavioral observations consistent with this interpretation include (1) larger-horned rams mate more often, (2) rams ritualistically brandish their horns when threatening other rams, and (3) rams prefer to hang out with rams having similarly sized horns (implying they visually discriminate horn sizes).

The evidence all points toward horns functioning as visual symbols of a male's fitness. But is this the primary function of horns? Big horns are incredibly powerful weapons. They act as substantial shields against attacks from predators. They also protect rams during head-butting battles for dominance and mating rights, which can last tens of hours. Observations of how sheep use their horns are critical to understanding their function. Similarly, observing how whales use their songs is crucial to understanding their primary function.

publicly available recording of a singing humpback whale. I'll describe the less obvious features of songs in the next three chapters. For now, knowing these three core properties of whale songs should suffice for you to start linking what whales sing to why they sing.

Echoic Consequences of Patterned Unit Production

It's time to invert the standard descriptions of humpback whale song. What happens when you adopt the perspective of an echolocating whale and focus on the sounds coming into a singer's head rather than the ones going out? A singer may hear all kinds of sounds when not bellowing, including the songs of other whales. If the goal of songs is to reveal whales' worlds, then most of those received sounds will function as the backdrop against which singers scan for behaviorally relevant echoes. Not all returning echoes are useful to singers. Listening singers in the breeding grounds likely sift through received echoes and non-echoes to find sounds signaling the arrival and movements of other whales, especially those of the opposite sex.

Your perception of recorded humpback whale songs is severely distorted relative to what a singer is likely to notice. Your "what" is not theirs. But whatever whales *perceive*, it's partly a reaction to what they *receive*. And what they receive is in the physical world. In principle, scientists could measure what whales hear, including the kinds of echoes produced by songs. So far, there are no reports linking song-generated echoes to the social scenarios happening around a singer. Some clues, however, help to reveal what the echoic scenes encountered by singing whales are like.

First, the repeating patterns of units that singing humpback whales produce will lead to similarly repeating streams of echoes.

This kind of repetition provides opportunities for singers to compare what they're currently receiving with what they received during recent iterations of the same unit patterns. When no whales enter within a singer's range of detection, or *active space*, the repetitive echoes coming from distant stationary objects should basically remain the same. The constancy of this echoic scene is analogous to a cork attached to a fishing line, which floats with predictable monotony when no fish is pulling on the hook. Each distinctive unit within a phrase provides a separate stream that a singer may monitor. Any large targets swimming into a singer's active space potentially introduce new variations in the returning echoes, analogous to the attention-grabbing cork movements that reveal a fish nibbling at the bait.

Now it's time to put your memory of chapter 4 to the test. The units that singing humpback whales produce within repeating phrases usually vary in pitch. What effect do you think these pitch differences have on the echo streams that the units generate? The short answer is that the distance to and depth of the reflective object dictate which units generate the strongest echoes. Some units within songs—those that remain audible from the longest distances—are more likely to generate detectable echoes from whales or other large objects located several kilometers away.

Studies of bat echolocation show that many species of bats produce longer duration cries focused on one to four pitches when they're searching and when they're farther from detected targets. Concentrating vocal effort on a small number of tonal components increases the propagation potential of a bat's cry because all the energy is channeled into a few waveforms,[16] somewhat like a laser

16. An acoustic waveform describes properties of a propagating sound wave: how rapidly particles are oscillating, how intensely, and how those oscillations vary over time.

light show. This also makes the cries and their echoes less susceptible to interference and easier for bats to hear amid background
noise. The same holds true for tonal units within humpback whale
songs. Longer-duration, fixed-pitch units are the ones within
whale songs that generally travel the farthest.

There's more to echolocating than just detecting the presence
or absence of targets. For echoes to be perceptually useful, a singer
needs to be able to track and predict the movements of other
whales. Research on bat echolocation has revealed that the best-
suited sounds for tracking targets are not the sounds that travel
the farthest distances. Instead, short sounds that rapidly change in
pitch are more useful for precisely determining changes in a target's movement patterns. This is why many species of bats switch
to using shorter, more slide-whistly cries when closing in on an
insect that might perform rapid, evasive maneuvers to escape
capture. Similarly, the units within humpback whale songs that
produce the echoes that best reveal other whales' movement patterns are not the units that travel the farthest. The units a singer
includes within each phrase thus determine not only the echoes
that the phrase is likely to generate but also the distances from
which those echoes are detectable, as well as the kinds of information that the returning echoes can potentially provide to the
singer.

Different kinds of units will yield different kinds of echoes. Only
a subset of units will best reveal the movements of another whale
swimming far from a singer. In the habitats where humpback
whales are most often heard singing, no single sound will generate strong echoes at all the relevant distances in all the different
water depths where a whale might swim. By morphing units along
multiple dimensions, singers may be exploring different focal
ranges, somewhat like optometrists do when they try out different
lenses on a patient to identify which one provides the clearest

image. In other words, whales may sweep through a series of acoustic "lenses" (the different unit morphs) to increase the chances that at least some of the units they produce will be optimal for generating informative echoes in the specific habitat in which they happen to be singing.

From the perspective of the sonar hypothesis, the main reason why humpback whales cycle through different themes while singing is that each theme offers different "looks" at the surrounding region, potentially providing unique perceptual information. For example, some of the themes that Scott and Roger described in their initial report contained mainly longer-duration units with minimal pitch changes—good for detecting faraway targets—while other themes contained more rapidly changing, shorter-duration units that are better for tracking moving targets. Still other themes contained both kinds of units.

As noted earlier, singing humpback whales typically vary the time they spend producing each theme, providing clues about how they may distribute their search effort. At a minimum, phrases that are repeated more times provide more potential "looks" at specific distances. By shifting the time spent producing each theme, singers control the amount of time they can listen for echoes generated by specific types of units. Individual singers vary the time they spend producing different themes throughout a song session. Songs are seldom repeated exactly, suggesting that each singer is controlling how it progresses through themes

Overleaf: *Tracking tiny targets in the dark.*
Echolocating bats hunting at night vocalize to detect and monitor the flight paths of insects as small as gnats. Humpback whales singing in light-limited ocean environments may likewise use echoes to apprehend the movements of distant "miniscule" whales.

on a song-by-song basis, possibly in response to the echoic feed-back received from each song.[17]

For the last 50 years, researchers have marveled at the complex structure evident within the songs of humpback whales. They've described the long, elaborate cyclical patterns that singers incessantly produce while hanging out along the coasts of tropical islands. Along the way researchers have come to believe that singing humpback whales memorize songs as a human musician might memorize a medley of popular sea shanties, rotating through the "greatest hits" in a fixed order to highlight their singing skills to any whales within earshot. Some view the patterning of units within songs as evidence that singers follow a kind of syntax that determines the order of units, similar to how the syntax of human languages determines word order. And of course, some listeners find whale songs to be intriguingly beautiful and compelling.

Contemplating whale songs along these lines runs the risk of transforming a whale's vocal actions into a behavioral snowflake. Like whale songs, snowflakes feature aesthetically compelling, ordered structures that are beautiful to behold. Like whale songs, the intricate structural complexities of snowflakes are bewildering in their variations. One can draw comparisons between the architectural details of snowflakes and the complex creations of the most sophisticated geometric artists. This approach will yield few insights into the principles underlying snowflake formation, however, and none regarding their function.

The patterns of units that singing humpback whales produce provide important clues about how and why they vocalize so intensively and for such impressively long periods. According to the

---

17. A note for number nerds—analyses of the distributions of theme durations produced by singing humpback whales show that these distributions are all similarly shaped. Intriguingly, the shape of these distributions matches that observed for the durations that raptors spend staring at different locations when searching for tiny targets, a convergence that will come up again in chapter 8.

sonar hypothesis, units are the fundamental functional element that singers are controlling, not because singing whales care about (or even perceive) the units they produce but because the properties of each unit determine what echoes a singer can potentially receive. Or, more precisely, the sequence of unit properties a singer produces determines what changes in echoic scenes a singer can potentially apprehend. It's these changes that may enable a singer to perceive the presence and trajectories of other whales swimming far beyond their visual detection range.

Knowing *how* whales produce unit sequences is thus important for understanding what whales are hearing and what they're attempting to do when they're singing. Are singing whales showing off their mnemonic and vocal prowess? Or are singers systematically searching? Are the mechanisms whales use to produce songs more like those of singing birds, dancing humans, or echolocating bats? In revealing how whales sing, chapter 6 will resolve these mysteries and reveal hidden physical mechanisms that may strongly affect how singers construct streams of incoming echoes.

*** *A Voice of Reason* ***

THE CASE FOR REJECTING THE SONAR HYPOTHESIS

You've made it halfway through the book! At this point you may be convinced, intrigued, or highly skeptical of the possibility that singing whales use their songs for long-range sonar. Regardless of your take, I thought it might be useful for you to hear someone other than me explain why whales sing and why most whale researchers don't think whales use their songs for echolocation. Toward that goal, this interlude provides the perspective of a dolphin sonar expert, Brian Branstetter, on the functions whale songs serve.

Like me, Brian started his journey into the world of marine mammal science by working in Lou Herman's lab, training the dolphins of Kewalo Basin to perform sophisticated tasks to reveal what's happening in their heads. Brian migrated from Lou's lab to the second most famous dolphin lab on Oahu, where he worked with Whitlow Au, before joining one of the oldest dolphin research facilities in the world, the US Navy Marine Mammal Program in San Diego. There he continued conducting experiments with bottlenose dolphins and orcas for over a decade to determine how they perceive their worlds using echolocation. Currently, Brian is helping with conservation-based research on the effects of anthropogenic noise on dolphins and whales at Naval Facilities Engineering Systems Command (NAVFAC) Pacific.

I reached out to Brian to seek a scientific perspective from someone who knows a lot about cetacean sonar but has no skin in the game when it comes to explaining why whales sing. He's contributed to several studies of humpback whales, including one that documented humpback whales balancing bottlenose dolphins on their heads[18] and another that explored hearing sensitivities in singing humpbacks. But whale songs have never been the focus of his research. In some ways, Brian's experience echoes Whit's. They both learned about biosonar by experimentally testing what dolphins can do, and they both (like me) became involved in studies of singing humpback whales somewhat indirectly.

Brian agreed to answer my questions from his home in Oahu. For me, it was a return to the land where it all began. My journey to Hawai'i 30 years ago was sparked by an intense interest in understanding what the minds of dolphins could achieve

---

18. As described in a paper by Mark Deakos, Brian Branstetter, and colleagues entitled "Two unusual interactions between a bottlenose dolphin (*Tursiops truncatus*) and a humpback whale (*Megaptera novaeangliae*) in Hawaiian waters"; published in the journal *Aquatic Mammals*, volume 36, pages 121–128.

cognitively. This time, I arrived (electronically) to gain insights into the mind of someone who's spent most of his life studying the sonar of dolphins. Below is a transcript of our conversation (edited for clarity).

Q: *How would you respond if your eight-year-old son asked you why whales sing?*

Branstetter: I would tell him we don't know for sure, but most people think they are singing to attract females. The reason we think this is because we see a lot of other animals that only sing during the mating season that are males and typically what they are doing—whether they're crickets or frogs—is trying to attract females. The song is sending some information to the females about how fit they are, whether they're big and strong, things like that. And that's what most people think humpback whales are doing because only the males are singing and they're only doing it during the breeding season.

Q: *What if he asked about dolphins singing?*

Branstetter: I guess it depends on your definition of song. Typically, people don't think that dolphins are singing. They have a wide repertoire of vocalizations. They produce whistles, they produce clicks, they produce squawks and squeaks. But scientists don't think dolphins are singing because people typically reserve the term song for some type of complex vocalization that repeats.

Q: *Is it reasonable to assume that when toothed whales produce click trains, they are echolocating?*

Branstetter: Oh, yeah, we know definitively that's what they do. You can put blindfolds on them, and they will track and capture a target.

Q: *What about when foraging humpback whales produce click trains, such as in the observations collected by Alison Stimpert?*

Branstetter: The clicks that Alison's published—I think they are probably echolocating. I really do. I don't know why they wouldn't use that for echolocation. It'd be nice to get more data on it. There's only been a few observations. If those clicks were used for some type of echolocation, it wouldn't surprise me at all.

Q: *What sorts of evidence would convince you that a singing humpback whale is vocalizing to generate useful echoes?*

Branstetter: What I would like to see is them approach a target and change their signal as a function of approaching the target. For most animals that echolocate, that's what you typically see. If you see a dolphin approaching a target, the inter-click interval is going to get smaller as a function of range to that target. Bats do the same thing. They'll even change the duration of their signal: when they are far away, they'll use a longer signal. When they get closer, the signal gets smaller and smaller. You see the same thing with beaked whales when they're swimming down. It seems like they're echolocating on the ocean floor. They have this long inter-click interval, but as soon as they detect a target, they switch their echolocation signals as a function of that target's range. I haven't seen anything like that with humpback whale song. That's probably the primary evidence that I haven't seen that I'd like to see.

Q: *When dolphins echolocate at long ranges, they sometimes switch to using these click packets, which I think you saw at San Diego. What do you think the dolphins accomplish by doing that? Why would they switch to using packets?*

Branstetter: We don't know. Basically, they'll send out a packet and the interval between the packets is also a function of range.

They're going to get more information. Maybe the echoes are going to be quieter. And because you're getting more of those echoes, the probability of you hearing a very quiet echo is going to increase, so you get more observations. If you send one click out, you're only going to get one echo, and you might miss it. But if you have four or five echoes coming back, that's four or five observations. Statistically speaking, you're going to have a higher probability of detecting a long-range target.[19] That's my best guess. I don't think the animals are integrating those echoes together to produce more energy or anything like that.

Q: *Why do you think many whale scientists assume that baleen whales only vocalize to socially interact or to convey emotions?*

Branstetter: I don't know if they do [assume that], but it seems likely they're doing it by analogy with other animals. When you see animals producing sounds, a lot of times they're just letting other animals know information about their emotions, their size, things like that. So, they're advertising information to other animals to communicate.

Q: *Have you ever had discussions with other scientists about the hypothesis that humpback whale songs could be a form of long-range sonar?*

Branstetter: We had a full seminar when I was in graduate school. Neil Frazer and Whitlow Au were there and there was a huge debate. I think Whitlow ended up driving Neil Frazer home afterwards. So, it was a very civil debate. The main arguments

---

19. I think this is an important insight—that long-distance targets generate quieter echoes and that extending the signal, say by repeating multiple clicks in a packet, increases the likelihood that some detectable echoes make it back. This is essentially my argument for why singing whales must use longer-duration signals (like bats) to detect echoes from targets at distances far beyond what dolphins can detect using clicks.

were kind of what I already mentioned. When most people are out in the field observing singing, it really looks like there's a male just sitting there and sending out information. Usually, when you see an animal echolocating, you see it doing some type of search pattern and it's moving towards the target. Whales are just sitting there sending out this long, complex song, which is also another argument against song being sonar. Typically, biosonar signals evolved over time to be optimal signals to help them accomplish the goal of detecting and finding targets. The complexity of humpback whale song, where it's constantly changing over time—it doesn't seem like it's optimized for detecting any type of targets. And you don't see humpback whale song changing as a function of range, which you see in almost all echolocating animals. I think those are the primary arguments that people use against the humpback whale sonar hypothesis.

I know that if a whale's out there vocalizing, it's probably getting a lot of information about its environment, such as how deep or vast the ocean is. If there's an animal in a shallow water inlet in Alaska, for example, it can probably tell acoustically from its vocalizations that it's in a small volume as compared to being out in the deep ocean, just like we can tell the difference between being in a basketball court versus being in a bathroom. These animals are smart. They're producing sounds all the time. They're probably getting a lot of information about their environment.

Q: *If you had to speculate, what do you think are the odds that singing whales use their songs as a form of active perception, beyond just sensing if they're in shallow water?*

Branstetter: I would say it's probably fairly high that they can do it, whether or not they're actually trying to do it. I'd say the odds are very high that they can do it just like a human can learn to echolocate if they need to. I bet you could train a humpback whale to do it.

If they had some reason to do it, I think the acoustics are there. They have the brain power to do it. Whether there's a need to do it, I don't know.

Clearly, Brian and I have different views on what the internal worlds (and needs) of whales are like. In contrast to many other experts in the field, however, he's all for me presenting my work on whale songs at scientific conferences. At the end of our interview, he encouraged me to attend an upcoming meeting of the Acoustical Society happening in Hawai'i to, in his words, "blow it up." Brian's skepticism about the sonar hypothesis captures the consensus view. Most researchers don't think songs could be used as sonar because songs seem too complex and because singers behave in ways that differ from echolocating toothed whales. And yet he acknowledges that if humans can learn to echolocate, humpback whales probably can too. And he interprets observations of humpbacks clicking while foraging in the same way he would interpret a toothed whale's behavior in similar circumstances—as evidence of echolocation. On this point we agree: Humpback whales should be capable of echoic perception.

CHAPTER 6

How Whales Sing

STAND FIVE FEET FROM A JACKHAMMER going full tilt and you'll get some sense of what it's like to be submerged near a singing whale. The experience is not simply sonic. You become a human vibrator. And yet aside from an occasional flapping of the flippers, which is often timed to the beat of the song, there is no visible evidence that a singer is doing anything—no movements of the mouth or head, no dramatic poses or posturing, no bubbles from the blowhole, few signs of swimming. Nothing. What you're likely to see in a video of a singing humpback whale is an animal that has mastered the mannequin challenge. The only clue that might indicate a humpback whale is singing is that, for reasons unknown, humpbacks usually sing with their heads pointed toward the seafloor.

How do whales generate such intense songs without showing any signs of effort or movement? Like most mammals, they rely on lung power to make sounds. Unlike most mammals, whales are constantly surrounded by liquid, which makes getting airborne sounds out into the world more challenging. Try yelling with your mouth full of water and you'll see what I mean. To create high-intensity sounds underwater, whales must use their bodies like a bell. Just as pealing church bells spread sound throughout a town, vibrating baleen whales recruit the water particles surrounding them to spread acoustic waves for miles around. Exactly how they do this remains a mystery.

Singing humpback whales can produce sounds lower in pitch than the deepest bass voice of a human singer. Since they are big animals, it's not surprising that they're able to beat human singers in the contest of "how low can you go."[1] More surprisingly, whales can also sing pitches an octave higher than the highest notes produced by professional sopranos, a feat that is simultaneously impressive and baffling. How does an animal the size of a whale sing notes that exceed the range of all human vocalists combined? The vocal range of singing humpbacks even exceeds that playable on a piano. Biologists' explanation for humpbacks' expansive range is as predictable as it is unsatisfying—if lots of males do it, especially in situations where females are likely to be choosing mates, then females must like it. If other types of singing whales don't hit the high notes, then it must be because females of those species don't like it. Case closed.

There is a twist to this story, however, that makes the "females like it" explanation a bit more difficult to swallow. Humpback whales don't always sing the highest- or lowest-pitched sounds they can make in any given year. Some years, singers produce songs that are full of high-pitched sounds. In other years, singers might not include any such sounds in their songs. If female preferences drove the evolution of extravagant vocal ranges in humpback whales, you would expect competing singers to consistently show off their full capabilities. Singers revealing only portions of their vocal prowess each year makes about as much sense as peacocks purposely coating their tail feathers in mud before showing them off to prospective mates.[2]

---

1. In general, the larger the sound source, the lower the sounds it can generate well—think tubas versus trumpets.

2. And collectively choosing different sets of tail feathers to cover with mud each year.

Consequently, whale researchers have been forced to change tack in their explanations of "why." Now the preferred hypothesis is that female whales prefer not vocal range per se, but song complexity, the implication being that a wider vocal range enables singers to produce more complex songs. Clearly, singers can flexibly produce lots of different sounds in a variety of combinations, but it's not so obvious that they need such a large vocal range to do this. After all, human musicians use a narrower pitch range to create songs that are even more complex than those sung by humpback whales. To answer why whale song spans so many octaves, we need to take a closer look at exactly how whales make sounds when they sing.

Bidirectional Blowing

Singing whales aren't the only animals that make sounds without moving their mouths or heads. Toothed whales can also do this. The invisible vocal acts of dolphins have frustrated researchers for decades; it's nearly impossible to keep track of who's saying what to whom when everyone's a ventriloquist. Luckily, dolphins can be trained to vocalize while an endoscope records the internal tissue movements that generate sound within their heads, providing a glimpse into how it's accomplished.

Dolphins make sounds with their noses. The part of a dolphin's nose that generates sound is hidden deep inside its blowhole. To visualize how dolphins make sounds, imagine that your lips have been split into two pairs and sucked into your nasal cavity[3] and that you can pinch your nostrils closed without using your fingers. Now blow your nose with your nostrils closed. A dolphin can voluntarily control how rapidly air is pushed to, and through, each pair of lips within its nose and can whistle using one pair of nose lips while producing streams of clicks with the other. This setup

----

3. One pair per nostril.

may even beat out the trunks of elephants for the title of weirdest nose in the animal kingdom.

You might expect that larger whales, having developed from a common ancestor within the same watery habitats, might possess similar ways of making sounds. You would be partly right. Baleen whales also voluntarily control when and how air flows through their noses when making sounds. But whales don't possess the "nose lips" sported by dolphins. Before the 1950s most scientists thought baleen whales were voiceless. The only reason that we currently know anything about the anatomical structures that humpback whales use to blast out their songs is because of the courageous effort of a singular scientist from New York City, Joy Reidenberg. Joy is an anatomist who has spent much of her life seeking out whale carcasses, crawling inside them, and carving out the humongous tubes that lead from a whale's lungs to its mouth and blowhole. For the past 30-plus years, she has collected these mega-tubes, measured them, and then stored them for later viewing by gaping scientists who have never seen a vocal organ that could potentially be used as a backpack.

The bizarreness of a humpback whale's vocal apparatus makes a dolphin's double-lipped nose seem almost ordinary. At first glance the vocal tract of a humpback appears similar to those found in most terrestrial mammals. It consists of a long tube (the trachea) that starts at the lungs and later splits into two separate tubes leading to the mouth and blowhole.[4] Within this otherwise standard configuration, whales possess two oddities. First, the membranes that vibrate to produce sound lie along one side of the trachea rather than across it; most animals that vocalize, including dolphins and humans,[5] do so by pushing air straight through vibrating

---

4. Or nose, in terrestrial mammals.

5. When you speak or sing, you are shoving air through a pair of tissue flaps in your throat, causing them to oscillate. These oscillating flaps create sound waves in the air column within your throat that propagate out through your mouth.

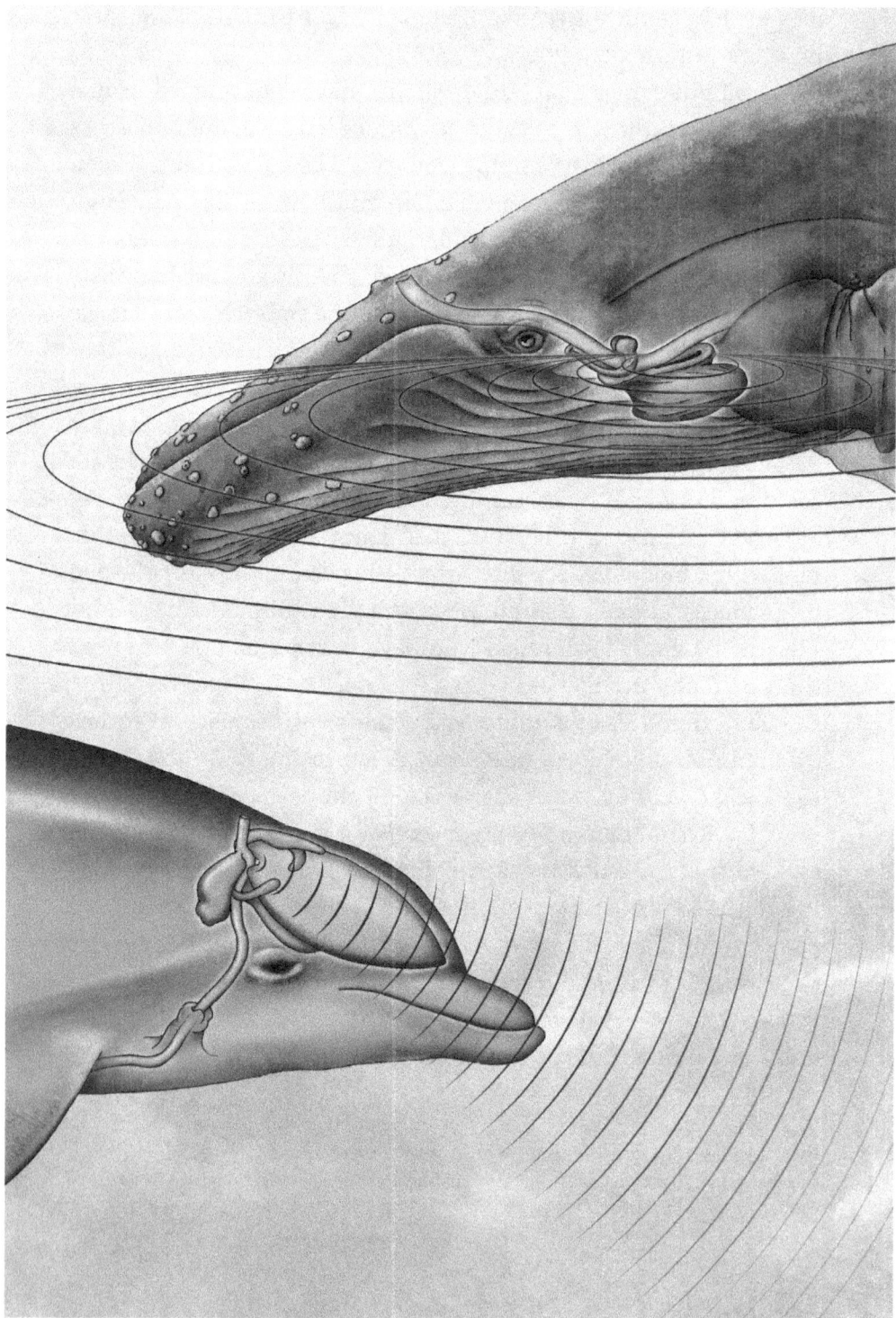

membranes rather than over them. Second, these membranes, described by Joy as *U-folds*, are positioned near the opening of a large, bag-shaped organ called the laryngeal sac. Dolphins have several small air sacs surrounding their nose lips, but none are as structurally complex as the humpback whale's laryngeal sac. Researchers think that humpback whales produce sound by passing air through the U-folds and into the sac, causing the folds to vibrate. Humpback whales have many muscles surrounding the sac and so should be able to push air out of it, as well.[6] This potentially gives them a third "lung" they can use to drive vocal membrane vibrations in the same way regular lungs do, but with air traveling in the opposite direction.

Humpback whales can sing underwater for 20 minutes or more without surfacing or releasing any air. This means that unlike terrestrial animals, the air stays inside their bodies when they make sounds. Given how loudly whales sing and how long, it seems unlikely that they're limited to emptying their lungs once while submerged. Consider what it would take for you to blow a party horn as loud as you can for 20 minutes.[7] You'd be hard-pressed to keep it up for more than a minute before needing to inhale more air, which is not an option when you're underwater. A more plausible scenario is that whales intermittently push air back into their lungs for reuse while singing. Singers may make sounds both when air

------

6. This sac may function not only as a vocal organ but also as a pneumatic valve that enables whales to rapidly change their trajectory while swimming.

7. This would still not come close to the sound volume that singing humpback whales produce.

Opposite page: *Cetacean sound sources.*
Singing humpback whales generate sounds by vibrating their vocal folds in ways that send sounds in all directions. Bottlenose dolphins generate sounds that are more directional using "nose lips."

Humpback singing

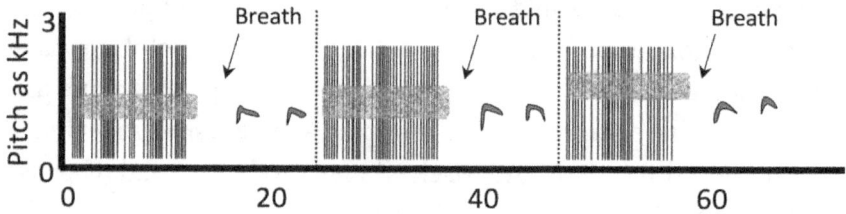

Human snoring

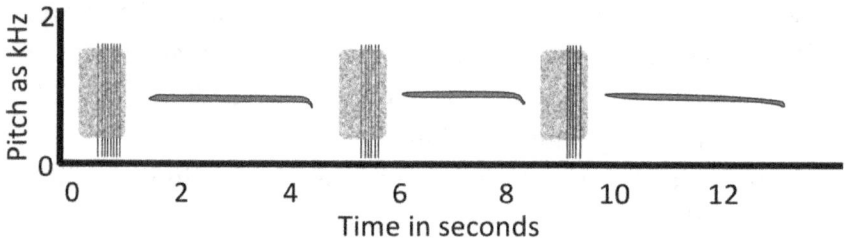

Time in seconds

Singing humpback whales appear to have the ability to make sounds when air is coming out of their lungs and when air is going back into their lungs. One way to gain clues about which sounds are produced in a particular direction is to follow the air. Surfacing singers often breathe rapidly between units. Sounds produced after an inhalation are probably created by air flowing from the lungs into the laryngeal sac. The air would then need to move back into the lungs before the next exhalation, creating a new opportunity for sound production. Comparative evidence showing how sounds produced by inspiration differ from those produced during outward airflow can be seen in recordings of snoring humans. When a snorer breathes in, the resulting sound is noisy (shown here as splotches). Snore sounds produced in the opposite direction are more like a whistle (*appearing as solid horizontal lines above*). Similar alternations in the units produced by a surfacing singer are consistent with bidirectional sound production.

travels from their lungs and when it returns to their lungs, as happens in hee-hawing donkeys. In this scenario, laryngeal sacs act as both a receptacle for air from the lungs and as a source for delivering air to the lungs. If singers produce sounds bidirectionally, they do it while controlling the number, duration, intensity, and quality of sounds generated when air flows in each direction, requiring a much higher degree of vocal control than donkeys possess. Such precise control of air movement rivals that shown by human musicians playing wind instruments.

The neuroanthropologist Terrence Deacon proposed that the evolution of voluntary respiratory control mechanisms in primates made it possible for early humans to develop their singing and speaking capabilities. Voluntary control, which depends on neural control of muscles, is what makes human sound production so flexible. You can independently control how tightly your vocal folds are stretched, how your tongue and lips are positioned, and how forcefully the air comes out of your lungs. Terrence argues that evolutionary increases in neural connections to vocal control mechanisms may explain why humans are the only primates with the ability to speak. Whales and dolphins must precisely control when they breathe, somewhat independently of when they make sounds, so they can avoid inhaling water. Undoubtedly, they, too, have developed extensive neural connections to vocal and respiratory control centers, which likely account for their flexibility in producing different kinds of sounds. Given this vocal flexibility, why aren't dolphins singing like whales or whales clicking like dolphins?

If researchers were restricted to field observations, it's possible that we would still know nothing about whether dolphins can sing. Luckily, a convergence of coincidences has revealed that dolphins are able to make some of the song sounds used by humpback whales. In 2008–2009, researchers at the Planète Sauvage dolphinarium in France were making recordings of dolphins whistling at night when they noticed occasional odd, low-pitched, "whale-like" sounds coming from the tanks. Looking back through their recordings, the researchers discovered that the dolphins started making these abnormal sounds soon after the park began broadcasting recordings of singing humpback whales to the public as part of their daily presentations. Dolphins are known to spontaneously imitate sounds they hear around them, and they are startlingly good mimics. I still remember the first time a dolphin imitated me as I whistled some random tune while walking past her tank.

The Paris recordings were the first time anyone had observed dolphins imitating a singing whale. The song sounds that the dolphins imitated were much lower in pitch than a dolphin can comfortably produce, so they transposed them up an octave. Even then the imitated song sounds were five octaves below the pitches of typical dolphin sounds. Notably, the dolphins only copied units from humpback songs, ignoring the timing and patterning of units within those songs. I know from personal experience that dolphins can easily match the timing and patterning of whistled melodies, so dolphins are not rhythmically challenged. It's more likely that the patterns within humpback whale songs are just too spread out in time for dolphins to recognize that there is a pattern.[8] If you play back whale songs, or even human singing, at slowed-down rates, you will see that it's harder to hear the rhythmic patterns. For a dolphin, even a single phrase within a whale song likely lasts an eternity.

Field recordings of humpback whales wearing suction-cup tags[9] recently revealed that humpbacks have a few vocal tricks of their own. It appears that humpback whales can produce trains of clicks with all the properties of those used by echolocating dolphins, with the exception that the whales' clicks are not ultrasonic.[10] This

---

8. Dolphin whistles are like words in that their properties can change multiple times within a second. In contrast, a single phrase within a humpback whale song can last 40 seconds or more, with seconds of silence between each unit within the phrase—slower than sloths speak in cartoons.

9. These "tags" are noninvasive data-recording devices that store any sounds produced by the wearer, the whale's movements in three dimensions, and any sounds that reach the whale. They provide a whale-centered view of a singer's experiences and actions.

10. These are the clicks discovered by Alison Stimpert and Whit Au that I asked Brian Branstetter about at the end of chapter 5.

finding came as quite a surprise to many whale biologists since the current dogma is that no baleen whale can echolocate. In fact, the researchers who discovered this were reluctant to even suggest the possibility that such sounds might serve any function comparable to echolocation, despite the obvious similarities to dolphin click trains. Perhaps, like the French dolphins, humpback whales are reproducing interesting sounds they hear in their habitat simply because they can. On the other hand, humpback whales produce these click trains in situations where they seem to be feeding, so they are probably not just playing around when clicking. It's unlikely humpbacks use clicks to detect individual fish like dolphins do.[11] What does an animal the size of a humpback whale really care about a single fish? Instead, humpbacks may click to perceive larger, more relevant objects like the seafloor, schools of seafood, or other whales.

If dolphins can make song(ish) sounds and humpback whales can produce click trains kind of like a dolphin, then why did these two groups evolve radically different voice boxes? The surprising answer may be that whales and dolphins developed different modes of sound production to achieve the *same* function but at very different temporal and spatial scales. As discussed in chapter 4, whale songs travel multiple kilometers, whereas dolphin click trains are rarely detectable from distances of more than a few hundred meters. When whales sing, their movements are quite limited. Dolphins, in contrast, are constantly in motion when they are either whistling or echolocating. The timing of sound patterns within whale songs is highly regular across minutes and even hours. The timing of click production by echolocating dolphins is dynamically matched to the rapidly varying distances of targets

---

11. Recall from chapter 4 that these kinds of clicks won't work as sonar over the kilometer distances that singing whales can potentially monitor using songs. They can work over shorter ranges, though!

with a precision of milliseconds. Dolphins travel and communicate within relatively tight-knit, familiar social groups, while humpbacks are usually strangers passing in the night. All these factors point to large differences in the spatial scales, social contexts, and acoustic environments within which whales and dolphins use sound. For whales and dolphins to use sound effectively in such disparate scenarios, they need different ways of controlling when and how they make sounds.

As noted earlier, singing whales can produce intense songs continuously for periods of 20 hours or more, across many months. No one has monitored the vocal output of an individual whale, but given that whales sing both day and night, it's possible they could spend the equivalent of a human's entire waking life singing. Perhaps the unique vocal adaptations seen in humpback whales were driven by the need to efficiently produce prolonged sessions of vocal patterns capable of ensonifying large swaths of ocean. Dolphins, meanwhile, evolved mechanisms that help them produce precisely timed clicks on the fly, capable of revealing detailed information about small objects located in one specific spot.

If singing whales go for maximum spatial extent rather than temporal precision, then this could account for why they use different sounds to echolocate, but it doesn't really explain why their songs are so structurally complex. Songs sound a lot fancier than dolphin click trains—to humans. The complexity of songs is one of the main reasons why whale researchers think that songs are acoustic peacock tails. Scientists do not perceive songs the way whales do, however. Complexity is in the ear of the beholder. The complex patterns that humans see and hear within songs may bear little relationship to what listening whales perceive, just like the complex structure of eye movements in humans watching movies reveals little about what they are seeing.

SEEKING SIMILAR SINGERS

Virtually every scientific paper or book discussing singing humpback whales at some point compares their vocal behavior to that of songbirds. This is partly because both groups produce complexly patterned sound sequences and partly because they both are thought to use those sequences for similar purposes. But when you get down to the details of what singing whales produce compared to what birds produce, very few birds do anything remotely similar. This is because humpback whales gradually change their songs as a group throughout their lives, while most adult birds do not.

One species of songbird that does collectively change their song in this way is the yellow-rumped cacique (pronounced "kuh-seek"). Some surface similarities are shared between these birds and singing whales. Like male humpback whales, male caciques promiscuously mate with females without forming long-term pair bonds. The male birds usually sing beside preferred nesting sites, directing their songs at nearby females and other males, but the singers do not defend fixed territories. The dominant males sing the most songs; younger males do not sing.

A male cacique sings five to eight distinctly different songs, each of which lasts about the same duration as one unit produced by a singing humpback whale. No note types are repeated either within or across these songs. All singers within a colony produce songs from this set, and as time goes by, all singers gradually change the properties of sounds within some of these songs in the same way so there is little variation between individuals within a colony. In contrast, clear song variations can be heard between different populations of yellow-rumped caciques.

Unlike humpback whale songs (or themes), many of the songs used by caciques are associated with specific social behaviors performed by males, like displacing a rival male or initiating flight. Also dissimilar to whales, singing in caciques is typically triggered by the appearance of a female. So although yellow-rumped caciques do collectively vary their songs over time, the songs they change differ radically in content, context, and function from the songs produced by singing humpback whales.

When Complexity Is Not So Complex

Much has been made of the complex structural features of humpback whale songs over the last 50 years. The apparent complexity of songs is what led humans to include whale songs on the famous Golden Record shipped off into outer space aboard *Voyagers* 1 and 2 and what inspired the writers of the more infamous *Star Trek IV: The Voyage Home* movie, in which the Vulcan Spock saves the world by helping humpbacks sing to alien invaders. One wonders, though, where the complexity of songs really comes from. Is it a product of sophisticated rules that singers use when constructing and producing songs? Or might the complexity of songs lie more in the eyes (and ears) of human observers? Could the apparent complexity of humpback whale songs arise from simple production processes, much like the complexity of snowflakes stems from the relatively simple effects of humidity and temperature on ice crystallization?

The cyclical structure of whale songs described in chapter 5 suggests that some unidentified factors strongly affect how singers progress through repeated patterns within songs. If whales produce sounds bidirectionally, then the regular features of song patterns may be partly caused by this two-way flow. Much like a creaking swing that generates regularly repeated sounds as long as a motivated child keeps pumping their legs, a singing whale may maintain a rhythmically steady stream of repeating phrases simply as the result of air recirculating during sound production. Unlike swings, whales' song patterns do change over time, though, and singers appear to have some control over how and when their patterns change. I say "some" because other factors may limit the kinds of patterns whales can produce. Specifically, submerged singers create hundreds of individual sounds using a fixed volume of air. As long as a singer stays submerged at the same depth, the total volume of available air inside their body will stay the same. The composition of gases within that air is not fixed, however. It gradually changes over time as physiological processes progress.

Singers must extract oxygen from the air in their lungs to stay alive, so over time the amount of oxygen decreases while the proportions of other gases, like carbon dioxide, increase. If you've ever sucked on a helium balloon before speaking, then you know that variations in gas composition can have dramatic effects on sound production. Might changes in the gas composition within a singing whale's body lead to similar variations in the sounds they sing? Increasing the ratio of carbon dioxide to oxygen in a resonating chamber has the opposite effect of adding helium—pitches drop. We don't really know which, if any, resonant air cavities help whales belt out their songs, but if resonance is important, then units should gradually decrease in pitch the longer a whale remains submerged. And sure enough, when recordings of songs are analyzed, the unit pitches gradually shift lower and lower until a singer surfaces, after which the unit pitch suddenly increases, followed by another gradual decline, and so on. This is exactly what we would expect to see in a resonating chamber where oxygen is cyclically depleted and then replenished.

Now you know of two separate physiological cycles that can physically affect the properties of sounds within whale songs. In one, air moves back and forth rhythmically inside the whale's head, hidden from view—the internal recirculation cycle. This recirculation cycle can affect the pacing of sound production as well as the kinds of sounds a singer makes in each segment of the cycle. For example, when you hear a donkey bray, the "hee" is always higher pitched than the "haw," and the "haw" always lasts longer than the "hee." These predictable patterns happen because a donkey's vocal folds vibrate differently when air is being inhaled versus exhaled. Whales' vocal mechanisms differ greatly from those of donkeys,' but their U-folds will still vibrate differently depending on the direction of airflow.

The second cycle is the singer's dive cycle, which typically consists of 10–20 minutes of submersion broken up by brief bouts of surfacing to avoid drowning.[12] Although we don't yet know how many times a singer recirculates air during each dive, if you assume that a singer resets after each repeated sound pattern, then the number of recirculation cycles per dive ranges from around 16–180, depending on how long the singer is down.

Beneath these seemingly simple processes of inhalation, exhalation, and surfacing lies a world of complexity. The two cycles may interact to create the equivalent of an acoustic double pendulum. A double pendulum is what happens when you take the bob at the end of a regular pendulum and replace it with a second smaller pendulum. Like breathing and internal air recirculation, the motions of a pendulum consist of simple back-and-forth movements. You might guess that a double pendulum moves pretty much the same as two pendulums sitting next to each other, or possibly in

––––––––––––––

12. Yes, whales can drown.

some other predictably repeating pattern. But the movements of double pendulums can be extremely complicated. In fact, the movements of this seemingly simple combo are so complex that mathematicians describe them as being chaotic—meaning that the present state determines future states, but those future states are effectively impossible to predict—like the so-called butterfly effect.[13] Both the double pendulum and the paired cycles of gas control in singing whales can lead to complex sequences of events that require no monitoring or control of the resulting structural features of those events.

Traditionally, whale researchers have described humpback whale songs as consisting of a set of distinct themes produced in a predictable order.[14] From this perspective, whale songs are like an acoustic carousel with themes taking on the roles of frozen prancing horses. Updating this model to better capture the possible effects of recirculatory and respiratory cycles requires replacing the fixed time it takes a carousel to complete a revolution with the more variable durations of dive cycles. It also necessitates swapping the steadfast up-and-down movements of the horses with something akin to horses sitting on springs, the motions of which depend in part on how the riders move and how fast the platform rotates.

A key piece missing from this modified merry-go-round metaphor is the operator that controls the ride. Once set in motion, the movements of a double pendulum are determined by friction and gravity. Unlike pendulums, however, singing whales show a high degree of control over the patterns they generate. Singers can start and stop singing. They can vary what patterns they produce, when they start each pattern, and how long they spend producing

---

13. A conceptual example in which the air movements caused by a single butterfly's flight can potentially affect the climate patterns on the other side of the planet.

14. At least most of the time—exceptions do occur.

patterns. They can also control how rapidly and in what direction they morph specific units within patterns, as well as their duration, timing, and repetition. The movements of air within a singer, and the singer within the water, are also under the continuous control of the whale. The physics of the two cycles can act together to shape the kinds of sounds that come out of a whale's head, but ultimately, it's the whale's brain that determines when a singer surfaces and what sounds it makes while submerged.[15] Far from being pendulous, singing whales constantly guide the dynamics of their cycles.

When it comes to hardcore pendulum-like sound production, few can compete with our familiar friends the plip-plopping bats who persistently alternate sonar signals like flying cuckoo clocks. I noted in chapter 5 that when I was a graduate student, I was startled to discover just how similar bats' alternating sonar signals are to the phrases produced by singing humpback whales. Now that you have a feel for how singing whales produce patterned sequences of units, you can better appreciate why seeing these similarities freaked me out. Some plip-plopping bats, like the big brown bat, alternate longer duration calls containing lots of different ultrasonic frequencies with fainter, shorter calls that contain a narrower range of frequencies. Other plip-ploppers alternate calls more rhythmically, switching between louder, longer calls centered at one pitch and fainter, shorter calls centered at a different pitch. Plip-plopping bats sometimes switch between producing patterns of faint-loud-faint or faint-loud-loud-faint, with the duration of silences following each call varying depending on whether the call is loud or faint. In short, plip-plopping bats vary multiple dimensions of their alternating calls and the gaps between them. They alternate which ultrasounds they produce, as well as their duration, intensity, and timing.

---

15. Through neural mechanisms that will be highlighted in chapter 9.

Singing humpback whales alternate their units in ways that are highly similar to plip-plopping bats. Singers alternate louder, longer duration units with fainter, shorter duration units and vary the durations of the gaps following units in ways that lead to predictable rhythms. If you slow down a plip-plopping bat until the pitches of the calls match the pitches of whale song units, the bats start sounding spookily like a singing whale. Of course, that's also true if you slow down some bird songs, as mentioned earlier, but birds are not alternating multiple acoustic dimensions in parallel like singing whales and plip-plopping bats.

Three decades ago, this vocal convergence of humpback whale song and bat sonar was surprising to me. At the time I had no reason to think that the ultrasonic calls of echolocating bats and the complex songs of whales should have anything in common. I had no inkling that echolocating bats were producing any rhythmically patterned calls, much less using alternating calls with acoustic relationships that matched those I saw within singing whales' phrases.

In hindsight, given that both whales and bats evolved their vocal capacities in contexts where maximizing the range of sound transmission was crucial, perhaps I shouldn't have been so surprised. On the other hand, since bats are not internally recirculating air or holding their breath while flying, it's telling that both groups nevertheless produce similar sound patterns. These similarities suggest that factors other than just physiological limitations drove the convergence of whales' and bats' vocal actions. According to the sonar hypothesis, the main factor likely to lead to such a convergence is the nature of the long-distance searching task that both groups face.

If you really want to understand how whales sing, you need to understand what's leading them to control their song production in the ways they do. Singers often remain motionless for long periods when singing. They spend a surprising amount of their time

Whale Song Acoustics: Songs as Sounds

It's not a given that every unit produced by a singing whale is intentional or serves some purpose. Some units, especially the quieter ones, may be incidental side effects of singing. Certain features of units that are less consistent than others might be "flourishes," sloppiness, or unintentional slips. For human singers, knowing what songs are supposed to sound like provides many clues about the skill of the performer. In the case of whales, researchers tend to treat all units within songs as being somewhat equal, although not all the features of those sounds are treated equally. Usually, those that are the most distinctive to human listeners get the attention.

In the case of human singers, the most noticeable song qualities are a combination of features the singer doesn't control, like the fact that male voices tend to be lower than female voices, and controllable features, such as whether the song is produced "in tune." A lot is known about how human singers' actions affect their vocalizations. Briefly, airflow, the tension of membranes in the throat, and tongue/mouth configurations are the main controllable features that humans use when singing. When the membranes (vocal folds) are relatively loose and floppy, the sounds produced by singers are quite choppy, sometimes called vocal fry, creaky voice, or popcorning. When the membranes are extra tense, the sounds produced are indistinguishable from a whistle—these ultrahigh-pitched sounds fall within what experts call the flute or whistle register. The modal register, the vocal mode normally used by singers, covers all the tensions between these two extremes.

Although humpback whales make sounds that differ in many ways from those of human singers, the units they produce map nicely onto the categories of sounds produced by vocalists. It's probable that whales, like humans, control what units sound like by varying the tension of vocal membranes. The biggest difference is that only skilled professional singers can produce the highest or lowest notes in human songs, and do so rarely, while all humpback singers appear to be able to produce the lowest- and highest-pitched units and do so relatively often. They are master membrane stretchers.

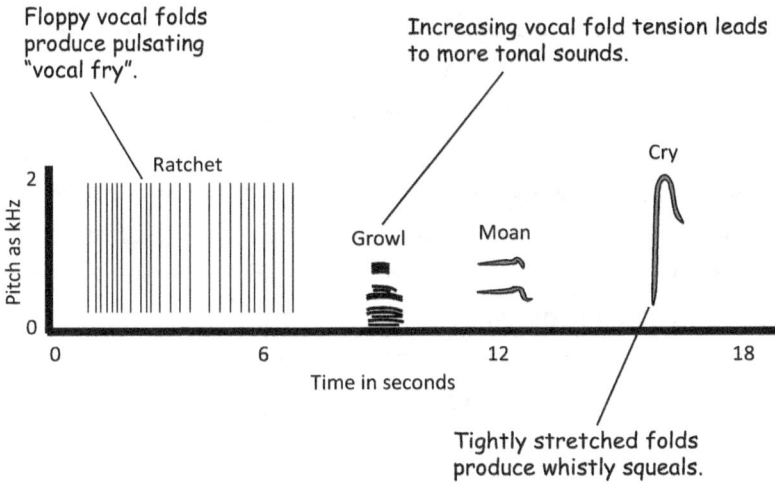

Floppy vocal folds produce pulsating "vocal fry".

Increasing vocal fold tension leads to more tonal sounds.

Ratchet

Cry

2

Pitch as kHz

Growl

Moan

0

0   6   12   18

Time in seconds

Tightly stretched folds produce whistly squeals.

The full range of vocal registers used by human singers maps almost directly onto the kinds of units present within whale songs, suggesting that singing whales also vary units by adjusting how their vocal folds vibrate.

singing in situations where there may be no whales close enough to hear them. Whales continue to sing even when surfacing, as if taking a break is not an option. And they precisely control and adjust the detailed acoustic features of the sound patterns they produce. Who or what are singers attempting to control through their incessant sonic cycling?

Proponents of the idea that whale songs are acoustic displays argue that singers are seeking to control the movements and decisions of other whales. The sonar hypothesis, in contrast, suggests that singers are searching for information that can help them decide how to control their own movements. Regardless of which of these two scenarios is true, whales make decisions based on what they hear during vocal performances. Chapter 7 will more closely evaluate what both singers and nonsingers hear during singing as a prelude to explaining how whales react to what they're hearing. Of particular interest are situations in which multiple singers can be heard producing overlapping songs, which detractors of the sonar hypothesis claim make it impossible for singers to perceive anything useful from the echoes their songs generate.

CHAPTER 7

Who Hears What

MARINE MAMMALOGISTS TRADITIONALLY define echolocation as "the ability to produce high-frequency clicks and to detect echoes that bounce off distant objects."[1] Thinking of echolocation in this way is a large part of why most marine mammal scientists discount the possibility that baleen whales can echolocate. From this perspective, without high-frequency clicks there is no echolocation. But there is nothing magical about ultrasonic clicks. High-frequency clicks are simply good for generating a certain class of echoes underwater—echoes that are especially useful for locating, tracking, and identifying small, fast-moving targets that are relatively close. If there's anything magical about echolocation, it's not the ability to click[2] or the ability to detect echoes. It's the capacity to construct mental images of the world by yelling at it. How one "yells" limits the kinds of images one might mentally construct from the resulting echoes, but it's how one listens that determines whether any image is constructed at all. This is why some blind people that grow up clicking can use echoes to navigate, while

---

1. According to Peter Tyack, per his chapter on "Functional aspects of cetacean communication" (p. 280), featured in the book *Cetacean Societies*, published in 2000.

2. A card passing over the spokes of a bike wheel can do that.

sighted bystanders just hear a bunch of clicks and perceive nothing else.

Consider what happens as you read. To an observer unfamiliar with reading, you may appear to be sedately contemplating an uninteresting rectangle. From your perspective, you're perceiving and interpreting sentences. An alien scientist might note that two fleshy orbs near the top of your body are darting about in synchrony in a semiregular pattern. That alien could define reading as the ability to produce rapid eye jerks and detect light that bounces off nearby objects. They wouldn't be wrong, but they clearly would not be capturing the essence of the experience of reading.

Understanding why whales sing requires understanding in a meaningful way what it's like for listening whales to perceive songs and song-generated echoes. Focusing on the sounds that whales make while singing and the contexts within which they make them can only get you so far. At some point you need to consider who's listening, what they hear, and what happens in their heads as they listen. That point is now!

All whales within a miles-wide cylinder surrounding a singer are potential listeners. Since singers are usually alone, most biologists assume that the relevant listeners are whales located a good distance away, including females who are considering the singer as a mating partner, as well as males judging the singer as a rival.

A few biologists think that humpback whale song is the equivalent of a bromantic frat chant that sends the message, "If you know the right tunes, you're one of us, and if you don't, then you should move along." From this perspective, songs are a communal display facilitating social interactions between males rather than a serenade to females.

In both the scenarios described above, the role of the listener is to assess songs and then decide how to behave in response. This requires a listening whale to compare incoming sounds to either some genetically encoded standard of excellence or to memories of previously heard songs. These comparisons need not be deliberate, voluntary, or conscious, but they must happen for a listener to select an appropriate response upon hearing the songs of a particular singer.

For songs to be maximally useful, listening whales must do more than just detect that songs are present. They need to track variations in the songs over time, compare those variations to some standard, and determine the spatial origin of the songs. These elements of listening are exactly those that an echolocating animal uses to construct auditory scenes from returning echoes. What differentiates using songs to communicate and using songs to echolocate are who the listener is and what they are listening for.

When it comes to listening for song-generated echoes, singers face a similar challenge as nonsinging listeners. Processing multiple streams of echoes is equivalent to separating songs from multiple audible singers, if those singers were all whispering. But the task of an echolocating singer is in many ways simpler than that of distant eavesdroppers. The singer knows the exact when, where, and what of the units it is listening for and can directly control what echoes arrive and when. The template an echolocating singer compares to incoming echo streams is the memory of the sounds he just produced, rather than memories of songs from months or years in the past. And those "working memories" are of the singers' own actions, which strengthens their fidelity, accuracy, and persistence. The singer clearly occupies the best seat in the house with respect to hearing what his[3] unadulterated songs sound like. Singers also have the most precise knowledge about the spatial origins of their song—no spatial hearing required. The main difference between a singer listening to self-generated echo streams and a distant listener is that nonsinging whales are presumably listening for cues to a singer's qualities. A whale using its

---

3. Or her.

songs for sonar, in contrast, would be listening for cues to non-singers' movements and intentions.

Sonic Self-Gratification

Many marine mammal biologists are skeptical that humpback whales use songs as sonar because they doubt singers can detect any echoes from their songs that bounce back from distant whales. In their view, "any echoes from another whale would be completely buried in noise."[4] The essence of this critique is that echoes from other whales are just too quiet for listening singers to hear.

It's certainly the case that echoes from distant targets can be very quiet, no matter the size of those targets. The US Navy, for example, is forced to use large numbers of sensitive hydrophones spread out over long distances and to broadcast sonar sounds much more intense than those produced by a singing humpback whale to detect and track submarines with sonar. As far as I know, no military has developed a sonar system capable of detecting whales using the kinds of sounds that singing whales produce. In that sense the use of songs as sonar is impossible . . . for human engineers.

Engineers and biologists concluded that humpback whales cannot hear echoes produced by their songs based on assumptions about how loud whales sing, what happens when a sound wave from a song hits a distant whale, the kinds of noise that listening whales must overcome to hear an echo, and the sensitivity of whale hearing.[5] Most importantly, they assumed that a whale's brain plays no role in determining the audibility of song-generated echoes.

---

4. From Whit and colleagues' paper, "Against the humpback whale sonar hypothesis," page 297.

5. Five years after making this argument, Whit and Au collaborated with a team of scientists to measure the intensity of the units produced by singing humpback whales and discovered that the units were significantly louder than originally thought. None of the other assumptions underlying this critique have been tested.

From their perspective, any faint echoes generated by songs are undetectable by a whale's ear, and therefore the whale's brain is irrelevant.

What these researchers failed to consider is that a listener's capacity to detect very quiet sounds depends heavily on how their brain processes auditory information, including the noise within which relevant sounds may be buried. Early researchers studying bat echolocation were amazed to find that broadcasting loud noises had almost no effect on bats' abilities to navigate through complex environments using echoes. Similarly, researchers studying deep-diving toothed whales are often at a loss to explain how they can detect echoes from animals like squid, which should be extremely faint. Echolocating animals regularly operate at the limits of what is physiologically possible, making any assumptions about what echolocators can or can't hear more likely to fall into that class of assumptions that make an ass out of you and me.

Determining what listening singers can hear is key to evaluating what information they can potentially gain from song-generated echoes. How does one figure out what whales can hear? Behavioral experiments provide the most definitive evidence of an animal's hearing sensitivities. Such studies are not currently feasible to conduct with baleen whales.[6] Physiological experiments, which involve recording neural responses to sounds, are another option. Some scientists have tried measuring auditory sensitivities in large whales by recording their brain activity with sensors placed on the whales' heads, often in nonideal contexts such as when a whale is immobilized within a net. The only other research avenue currently available is to closely examine an animal's ears and the

---

6. Typical experimental studies of hearing sensitivities involve training an animal to expect a certain sound to predict some relevant event and then varying the pitch and intensity of the sound until the animal shows no reliable reaction when the alerting sound is presented.

associated neural circuits. The observed anatomical structures can then be compared to those present in other species whose hearing capacities are known from behavioral and physiological tests.

Describing the ears of whales and dolphins is easier said than done. Outwardly, they appear to have no ears at all.[7] Internally, a slew of anatomical oddities make understanding cetacean hearing a bit of a nightmare. All the parts present in human ears are also present in whale ears, but in modified forms. For example, the human eardrum is a small, thin, circular membrane, shaped more like a cymbal than a drum. The eardrum in a baleen whale is a long, hollow sleeve of tissue, shaped somewhat like what you would get if you cut one finger from a rubber glove. In humans the eardrum converts air vibrations into tissue vibrations that trigger a chain reaction of movements in tiny bones, which ultimately enable you to hear. Scientists are not sure what's happening with baleen whale eardrums. They're not even sure if their eardrums still contribute to hearing. Weirdly, whales' eardrums appear to molt each year! Some scientists think that the ear canal of whales, within which the eardrum extends, may have been repurposed to detect the changes in pressure that occur as whales dive.

The uncertainty about baleen whales' finger-shaped eardrums comes from the fact that sound reaches whales through water rather than air. Terrestrial mammals sport relatively obvious fleshy funnels—the outer ears—that channel air directly into their eardrums. Your head effectively reflects most airborne sounds, leaving only the two tubes—your ear canals—as conduits of sound into your head. Not so for whales. The soft tissues and bones in their heads and bodies offer some resistance to waterborne sounds,[8] but many sound waves can travel directly into a baleen

---

7. Aristotle noted that dolphins can hear but found no evidence that they had ears.

8. The structures inside whales that reflect sound the most are pockets of air rather than bone or tissue. Consequently, it's the volume of air within a whale's lungs and other larger cavities that determines what echoes a whale's body produces.

whale's body without needing to pass through any specialized portals. That's because the soft tissues in a whale's body better match the physical features of salt water. Sound waves just keep going when they reach the water-to-skin boundary and only begin changing how they propagate when they run into oils, bones, and air pockets. Effectively, a whale's internal anatomy plays the role that your outer ears do with respect to channeling sounds toward the inner ears.

Scientists remain unsure of exactly how sound waves in water travel through a whale's body to generate auditory sensations. Some argue that sounds travel directly through tissues in the head to reach the inner ears, while others suggest that bones in the skull steer sound waves toward the ears. Early experiments on dolphins provided support for both routes of sound travel. In one experiment, speakers placed near a dolphin's lower jaws led to the strongest internal vibrations near the ears. Subsequent experiments showed, however, that this was only true for ultrasonic sounds and that lower-pitched sounds produced stronger vibrations when placed on the sides of a dolphin's head (in the positions where a terrestrial mammal's ear canals would normally be found). At least four channels within a dolphin's head seem to guide sound toward their ears: two along the left and right lower jaw bones and two on the left and right sides of the head straight out from the ears.

Researchers have just begun to examine how sounds travel through the heads of baleen whales. The skulls of baleen whales differ significantly in shape and structure from those of dolphins and other toothed whales, leading many scientists to argue that baleen whales failed to evolve echolocation capacities. Most of the unique features seen in toothed whale skulls, however, are specializations for producing and perceiving ultrasound. As I've noted earlier, ultrasonic clicks are not the only sounds that generate echoes, so these differences in skull shape only show that baleen whales did not evolve to echolocate using ultrasonic frequencies.

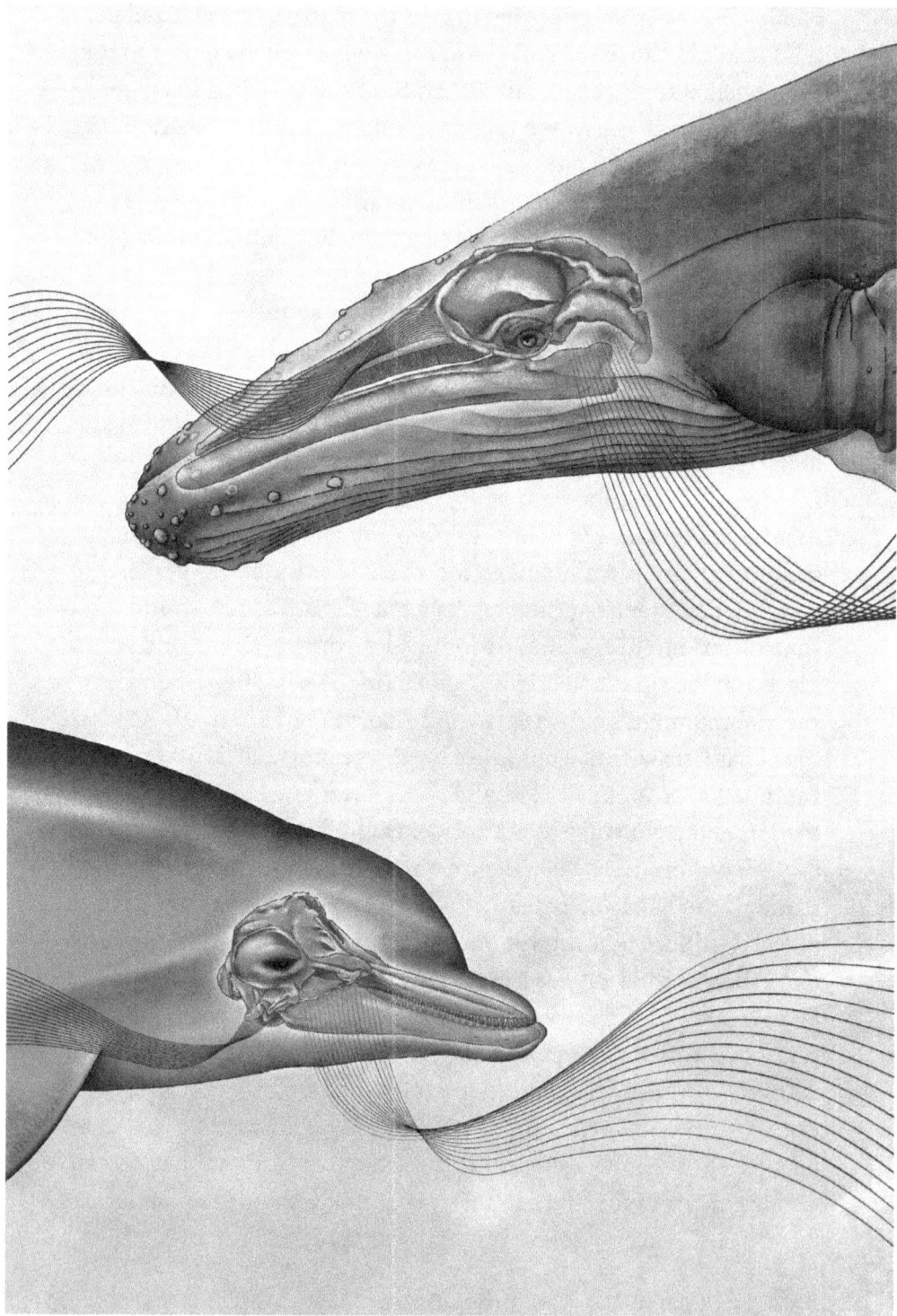

Early theoretical evaluations of baleen whale skulls hinted that they might act as a *waveguide*—a structure that channels the transmission of vibrations in a particular direction—similar to what researchers think happens in a dolphin's lower jaws when hit by a wave of ultrasound. But for baleen whales, the bones in their heads might enhance the reception of audible or possibly infrasonic sounds, rather than ultrasound. One sign that bone conduction may play a different role in baleen whales than in toothed whales is that the ear bones of baleen whales are fused to their skulls, such that skull vibrations may more easily transfer directly from the skull to the ear. Recently, computer simulations built using X-rays of the skull of a fin whale calf revealed that sounds appear to travel both along the jaw bones and through the flesh on the sides of the head. Although baleen whales' heads are structured differently from those of toothed whales, they, too, appear to have evolved multiple pathways through which sounds in water are transmitted to their inner ears.

Importantly, sound transmission through bone is faster than sound transmission through tissue, which is in turn faster than sound transmission through air. This means that as sound waves reach a whale's inner ears through different channels, they arrive at different times, with sounds traveling through bone reaching the finish line (the inner ear) first. This is also true for human hearing, but in air the number of paths through which sounds can reach your ears are mainly determined by the external shape of your head and outer ears. Whales have more channels to deal with since sound waves can approach their inner ears from all directions.

Opposite page: *Pathways to audition.*
Underwater sounds can reach cetaceans' ears by traveling directly through their bodies or may be channeled by skeletal structures. Both whales and dolphins seem to hear vocalizations through multiple internal pathways.

This might seem to put whales at a disadvantage, but differences in the arrival times of sounds provide listeners with important clues as to where sounds come from. Listening whales receive more spatial cues than terrestrial mammals and in principle could combine those cues to pinpoint the origins of incoming sounds.

A major problem occurs, however, when localizing sounds underwater because sound travels faster in water than in air. Your ability to hear where sounds come from depends on your ability to unconsciously detect tiny differences in (1) how loud the sounds are at each ear, (2) when the waves arrive at each ear, and (3) the pitches contained in the sound waves. Those differences are a result of what sound waves do when they hit your head. How sounds bounce off your head depends a lot on whether the sounds are high-pitched or low-pitched. Generally, you will be worse at figuring out where low-pitched sounds come from because there will be fewer and less detectable differences between the sound waves arriving at each of your two ears. When sound waves travel faster, as they do in water, bone, and blubber, there is less time between when they reach each of a whale's ears, making higher-pitched sounds in water more like low-pitched sounds in air. The worst-case scenario for a terrestrial animal like you attempting to localize sounds is a low-pitched sound produced underwater. Humans are horrible at judging where sounds are coming from when underwater. Extrapolating from human hearing, all evidence points toward baleen whales having essentially no ability to localize any vocalizations they hear, except perhaps (ironically) the ultrasonic echolocation sounds produced by toothed whales.

We know from behavioral observations that baleen whales have no problem localizing vocalizations produced by other whales. The fact that our theories of spatial hearing, derived mainly from studies of terrestrial mammals, cannot explain how baleen whales localize low-pitched sounds in no way prevents them from doing so. Humpback whales will rapidly swim directly toward playbacks of vocalizing humpbacks or toward groups of vocalizing whales from miles away. Baleen whales will also swim directly away from broadcasts of orca calls. Neither behavior is possible if whales don't know where the voices are coming from.

How baleen whales localize sounds remains a mystery, though it's generally assumed that they do so by combining signals from their two ears. That would mean whales have some internal mechanism for detecting subtle differences in the timing or properties of sounds reaching each ear. It's also possible, however, that baleen whales have systems that enable them to localize sounds in ways that terrestrial animals cannot. For instance, recent experiments show that the movements of water particles caused by a singing whale are detectable from longer distances than researchers expected. Fish have evolved sensory systems capable of detecting such movements when they are generated from nearby sources. It's possible whales have also developed unique sensory modalities that enable them to localize the origins of water particle movements from much longer distances.

Such mechanisms might enable listening whales to point their heads in the direction of a distant singer, but they would not explain how whales can judge the distance that songs, or echoes from songs, have traveled. My early simulations of sound transmission, discussed in chapter 4, suggested that humpback whales might be able to figure out how far away a singer is by listening for which pitches are absent from a received song. As songs travel long distances, different pitches die out at different rates, so if you compare the pitches that are present and those that are missing, you could potentially tell how far a song has propagated.

The main problem with using variations in pitches to judge how far a song has traveled is that the specific ways that propagation degrades pitches vary from one environment to the next. A whale would need to use different pitch clues to distance in various environments and would need to be familiar with an environment to judge distances to a sound source there. This might seem to rule out the possibility that whales compare pitches to judge singers' distances, but humans can only judge the distance to a sound source accurately when they are in familiar environments, so the same could be true for whales. In fact, humans also need to be familiar with the sounds they are hearing to figure out the distance to their source. If you are judging the distance to a speaker broadcasting speech in a familiar environment, you'll do okay. But if you have to judge the distance when that same speech is played backward, you'll be horrible at it. If you were to practice judging the reversed speech with some feedback about your accuracy, over time you'd eventually get better at hearing differences in the distances of backward talk.

Findings like these show that humans learn to hear the distances to different sound sources. You're not born being able to tell whether a barking dog is near or far. While it's true that sounds tend to get quieter the farther away you are from their source, such differences only provide very rough estimates of a vocalizer's distance. In thinking about the different ways in which whales, humans, and other animals might hear how far away a singer is, I began to look at how songbirds judge their distance from one another, specifically focusing on the songs of the tiny chickadee.

What I discovered was that chickadee songs seem to be structured in ways that defeat most of the mechanisms that terrestrial animals use to judge a vocalizer's distance. And yet chickadees use their songs in contexts where listeners need to know their distance from a singer, such as in social interactions related to establishing territorial boundaries. Surprisingly, the cue from chickadee songs that most reliably indicates the distance a song has traveled is a combination of reverberation generated by the first note in the song and the intensity of the second note in the song relative to the level of reverberation. I mentioned in chapter 4 that reverberation is persistent environmental echoes—essentially

HIDDEN CLUES IN SIMPLE SONGS

Like whales, birds sometimes use songs to judge their distance from a singer. This is particularly relevant for establishing and defending territories: A male singing 80 meters away might not be an immediate concern, while one singing 40 meters away may have crossed a territorial boundary. Similarly, a male singing to advertise the extent of his domain can only do so if other males can perceive how far they are from the singer.

The complex songs of birds like caciques and canaries provide lots of different cues that can reveal a singer's distance. The situation is more challenging, however, for birds like the black-capped chickadee. Chickadee "fee-bee" songs are some of the simplest produced by any animal, consisting of just two tones—basically the ding-dong announcing an elevator has arrived. Chickadees use their fee-bee songs in many different contexts, including during territorial interactions. Most of the ways that birds judge a singer's distance do not work on fee-bee songs. It's still a mystery how any listeners localize them. One intriguing possibility is that chickadees localize singers using reverberation.

Playback studies revealed that when the "bee" arrives at a listener, reverberation from the initial "fee" is still present in the environment. Surprisingly, how loudly the "fee" reverberates relative to the loudness of the "bee" provides a precise indication of the singer's distance. One unique feature of the chickadee's song is that singers often transpose their fee-bees into different keys, meaning they lower or raise both pitches at the same time in ways that maintain the melody. By transposing their songs, chickadees preserve the reverberation-related cues that can reveal a singer's distance.

Believe it or not, singing humpback and blue whales produce phrases that closely match the properties of the chickadee fee-bee song and transpose those phrases in much the same way as chickadees do (although at much lower pitches). In this respect there is much stronger convergence between the songs of chickadees and whales than between whales and birds that produce more complex songs. Understanding how chickadees perceive their reverberating songs and why chickadees transpose their songs can provide important clues about why singing whales produce similar patterns.

what happens when lots of echoes from many different sources overlap in time. In the case of singing chickadees, the production of a single note can generate echoes from many different surfaces (the ground, tree trunks, branches, etc.). Researchers tend to think of this mish-mash of echoes as ambient noise. But hidden within the chaos are signs of a singing chickadee's location.

By the time I began marveling at the unique ways that chickadees use songs to localize each other, I had pretty much abandoned studying humpback whale songs. My scientific research had turned more toward how people's brains change when they learn to make subtle distinctions between novel sounds, like speech broadcast from different distances, and whether ways might exist to alter people's brains more rapidly, a topic I'll explore more fully in chapter 9.

For that reason, I was caught off guard when a group from halfway around the world invited me to give a talk on humpback whale song. A conference on humpback whales was being organized on a small island off the coast of Madagascar, and the organizers wanted me to give a talk at the meeting. I was at first reluctant, given that I hadn't focused much attention on whale songs in several years. But lots of findings from my earlier research had never made it to print. Plus, I heard that my graduate advisor, Lou Herman, whom I hadn't seen in over a decade, would be attending, so I eventually decided it was worth making one last foray into the world of whales.

The conference, dubbed the Humpback Whale World Congress, included an eclectic mix of senior and junior researchers, all studying some aspect of humpback whale behavior, biology, or ecology. The meeting was the first of its kind, and it was exciting to see the different ways scientists were attempting to understand what humpbacks were up to. While attending the conference, I discovered that Lou was working on a new paper reviewing the possible functions of humpback whale song.[9] At the meeting he asked me to read a draft of his paper as an informal reviewer. I was not surprised to see that Lou made no mention of the possibility that songs could be a form of sonar in his review paper. From his perspective, that hypothesis was dead and buried.

Perhaps Lou's utter disregard for an idea that I'd spent so much time evaluating in his lab reignited a spark of indignation. Or maybe seeing other attendees persist in accepting explanations of singers' behavior that I viewed as totally inadequate awoke some feelings of scientific duty. Whatever the reason, I found myself once again thinking more deeply about what was going on with humpback whale songs. And then I had a second epiphany regarding what singers might be doing with songs: If chickadees could produce songs that reverberated in ways that might reveal their locations to listeners,[10] why couldn't humpback whales?

---

9. A paper that would ultimately be Lou's last scientific contribution.
10. If you skipped over the box in this chapter on "Hidden Clues in Simple Songs," now's the time to go back and check it out so that you know how song-generated reverberation can potentially help chickadees localize singers.

At first, I thought this idea was ridiculous. After all, I'd spent many years performing detailed analyses of humpback whale songs, as had many other scientists. Surely if something like that were happening, I or someone else would have noticed it by now. But then I realized that just as many scientists had analyzed chickadee songs without reporting anything noteworthy about the reverberation they produced. Maybe the same could be true for analyses of whale songs? When I started looking at recordings from humpback whale singers made not too far from Madagascar, I saw it—the reverberation that was hidden in plain sight, produced in patterns almost identical to those I had seen in the chickadees' songs. The singing whales were using much lower notes over much longer durations than the chickadees, and in more complex combinations, but in most other respects singing humpbacks were fee-beeing.

Why would singing humpback whales produce phrases that reverberate in ways that facilitate the ability of listeners to judge their distance? Among lots of possible reasons, the one that jumped to my mind when I began to see the pervasiveness of song-generated reverberation was this: If listening whales can precisely estimate the distance to a singer using song-generated echoes, then singers can also precisely estimate their distance from any large-bodied listeners using those same echoes. In other words, if humpback whales sing in ways that generate reverberation of a type that facilitates the spatial localization of song sources, then those benefits would apply to all sources of songs, including any nonvocalizing whales whose bodies reflect songs back to their source.

A "Whale's Ear View" of Whale Songs

Scientists typically describe cetacean echolocation as a kind of acoustic ping-pong in which the echolocator repeatedly clicks and then listens for any echoes from that click. The echolocator clicks, then listens, clicks, listens, and so on, with each "game" lasting one or two seconds and consisting of tens to hundreds of click-echo volleys. Echolocating dolphins integrate the information across all the echoes they receive to perceive what's happening in front of them. Many bats also vocalize in volleys when they echolocate, as do the few humans who echolocate, with the main difference between bat and dolphin sonar being that bats use cries instead of clicks and modify features of their cries based on how far they are from targets of interest. Bats and toothed whales also both adjust how rapidly they vocalize to account for their distances from targets.

This then is the canonical mode of echolocation for dolphins and most bats. In this mode, reverberation plays no role, other than perhaps as a kind of background buzzing that individuals must ignore or overcome to detect faint echoes from edible targets. Timing plays a key role because the interval between when a click is sent out and when its echoes return provides important information about a target's distance. In this mode, reverberation is the enemy because the more reverberation there is, the harder it will be for the echolocator to hear any echoes from targets.

For most echolocating bats and dolphins, reverberation is like a rising fog—for most, but not all. A select group of bats, called the horseshoe bats, evolved an alternative echolocation method. Horseshoe bats get their name from the fact that if you look at their faces up close, they have a flap of skin around their noses resembling a horseshoe. This skin flap is thought to affect how their cries propagate. Unlike many bat species, horseshoe bats often hunt their prey while hanging from perches. A horseshoe bat will hang from a twig producing cries at a fixed rate for many minutes before flying out, grabbing an insect, and then returning to a nearby perch. Their cries are focused steadily at one pitch, as if you repetitively pressed a piano key every time the second hand of a clock moved.[11]

Horseshoe bats typically hunt in forested areas. These bats are thought to have evolved their unique style of echolocating as a way of dealing with the high levels of reverberation caused by dense foliage. In effect, horseshoe bats generate a continuous stream of reverberation at a single frequency, and when an insect flies close enough to a bat, the insect interrupts this ongoing stream of background echoes, alerting the bat to its presence. This type of echolocation is specialized for detecting insects' fluttering wing movements. If a moth could glide past a horseshoe bat without moving its wings, it would become undetectable. A moth's best defense against these bats is probably to land or fall.

The tiny changes in echo streams caused by an insect's wing beats are exceedingly difficult to detect. They consist of subtle variations in the pitch of the returning echo stream known as a Doppler shift. Doppler shifts happen when the source of a sound is moving rapidly, like the sound changes produced by a race car as it speeds past. Since moths are not race cars, the changes in echoes created by their wings racing back and forth are miniscule. Other bat species cannot detect them. Horseshoe bats can detect these faint warbles because they focus all their vocal energy at a single pitch[12] and because their ears are specialized for hearing tiny variations centered around that pitch.

Sensory neurons within the cochlea (pronounced cock-lee-uh) are the main components within the inner ear that determine which sounds are detectable. The cochlea is a fluid-filled spiral structure that looks a bit like a snail shell. Movements of fluid

---

11. The bats actually produce about 10 ultrasonic cries per second, with intervals of silence between them that last about as long as the cries.

12. The acoustic equivalent of a laser—lasers focus energy at a single wavelength of light or within a very narrow range of wavelengths, and horseshoe bats focus energy within a narrow range of wavelengths of sound.

within this "shell" cause sensory receptors to activate, ultimately producing sensations of sound. In a horseshoe bat's ear, the position along their cochlea where echoes generate activity is more extensive than in other mammals and especially sensitive to the focal pitch produced by the bat. This specialized region is called the *acoustic fovea* (sounds like foe-vee-uh), in correlation to the fovea in your retina, which is the region where your visual acuity is highest. The overall number of sensory neurons in a bat's ear is similar to what's seen in other small mammals, about 16,000. You can estimate how sensitive a mammal is to a particular pitch based on the number of sensory neurons devoted to detecting that pitch. About a fourth of the sensory neurons in a horseshoe bat's ear are devoted to processing their echolocation pitch. This is a much greater proportion of neurons devoted to a specific pitch than is seen in humans and most other non-echolocating mammals. The resulting acoustic fovea provides heightened auditory sensitivity to that particular pitch.

This raises the question of what cetaceans' inner ears look like. Do they also have specializations for detecting subtle differences in faint echoes? Or are their inner ears more like those of humans, showing good detection and resolution for a broad range of pitches?

Most toothed whales use clicks to echolocate. Clicks are more like the sound from hitting a drum than hitting a piano key, so you wouldn't expect toothed whales to have an acoustic fovea tuned to a single pitch like that of horseshoe bats. But dolphins and other toothed whales do need to make subtle distinctions between faint echoes.

Neuroanatomical measures of toothed-whale cochleae[13] reveal that bottlenose dolphins have about 100,000 sensory neurons

---

13. Weird Latin plural of cochlea, pronounced "cock-lee-eye" because "cochleas" would be too easy.

innervating their inner ears, and belugas have about 150,000. That's approximately five times the number of sensory neurons in human ears. Of course, belugas are substantially bigger than humans, as are their ears, so one could argue that it's the number of neurons per millimeter of ear that matters more than the absolute number. Even then, humans have about 1,000 neurons devoted to each cochlear millimeter, while belugas have about 3,600 neurons per millimeter. Horseshoe bats come in at around 1,800 neurons per millimeter within their acoustic foveae. Since the neuronal numbers for toothed-whale ears are more than double the number seen in land mammals, neuroscientists have concluded that toothed whales need these extra neurons to resolve details of ultrasonic echoes when echolocating.

When neuroanatomists performed similar measurements of humpback whale cochleae, they discovered something interesting. The humpback whale inner ear contains about 160,000 sensory neurons, with close to 2,400 neurons per millimeter! Humpback whales monitor each millimeter of their cochlea with more than double the number of neurons seen in the human ear and collectively have more than five times the total number of sensory neurons to respond to incoming sounds as you do. Although it's not entirely clear how a humpback whale's hearing range compares to a human's, most evidence suggests that the overall range of detectable pitches is similar—much more similar than either species' hearing is to the hearing range of a dolphin or a bat.

Horseshoe bats do amazing things echoically with a pair of acoustic foveae that cover only portions of their cochleae. In terms of numbers of neurons, the cochleae of humpback whales are what you'd get if you stitched together about 90 acoustic foveae from a horseshoe bat. Humpback whales devote more neurons per cochlear millimeter to processing pitches than the most sensitive bats, however. Assuming a humpback whale's sensory neurons are

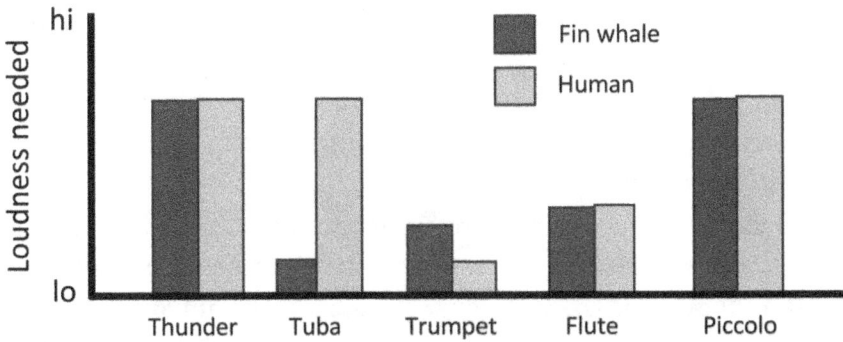

Current estimates of underwater hearing sensitivities based on behavioral tests in humans and on measurements of skull and tissue properties in fin whales suggest that the range of sounds that humans and baleen whales hear overlaps quite a bit. Both species seem to be best able to detect sounds in the ranges produced by musical instruments, which for humans are the same pitches they are most sensitive to in air. The lowest- and highest-pitched sounds that both species can hear need to be a lot louder for individuals to detect them. Oddly, estimates so far suggest that the underwater hearing of baleen whales is no better than the underwater hearing of humans and that humans are more sensitive to the lower-pitched sounds than are whales, though most baleen whale vocalizations are much lower in pitch than the sounds humans make. It's unlikely that humans can detect whale sounds underwater better than whales. Researchers will need to collect more direct measures of hearing in whales to determine which species is the superior listener when it comes to perceiving whale songs.

evenly distributed along their cochleae,[14] humpbacks have the equivalent of about 65 ultra-high-resolution acoustic foveae to work with.

The first time I saw these neuroanatomical descriptions of a humpback whale's inner ears, I couldn't believe it. Humpbacks have more sensory neurons innervating their inner ears than a bottlenose dolphin or a beluga? Their ears are more densely innervated

14. Anatomical studies of the inner ears of large whales are logistically challenging. As a result, many of the detailed neural features that have been measured in bats' inner ears have yet to be measured in any baleen whale.

than those of any echolocating bat? To explain this massive deploy-
ment of sensory neurons in humpback whale ears, researchers
have concluded that baleen whales must be capable of auditory
processing as complex as that of echolocating toothed whales. They
remain uncertain as to why.

No one knows to what extent humpback whales can detect faint
echoes or resolve subtle differences in the pitches they hear. What's
certain, though, is that humpbacks possess neural hardware that
puts them at the apex of the animal kingdom when it comes to
sensing sound, along with many other cetacean species. The sonar
hypothesis provides a straightforward explanation for why this is
the case.

Although humpback whales appear to have auditory sensing
power comparable to or exceeding that of a horseshoe bat, they
would certainly be unable to detect anything like the echoes from
moths' fluttering wings. The sounds humpback whales produce are
too low-pitched by a few orders of magnitude to reflect any echoes
from moth-sized objects. The wavelengths of low-pitched sounds
are too long to be affected by something as tiny as a moth. Similarly,
singing humpback whales would be unable to detect any Doppler
shifts created by other whales' movements because the resulting
pitch differences are way too tiny for any biological system to
detect.[15] The high degree of innervation within a humpback whales'
cochleae suggests that they should be able to make exceedingly precise distinctions between the pitches of incoming sounds, but no
one knows what this ability enables them to do perceptually.

---

15. You can hear increases and decreases in the pitch of an approaching or receding
race car (the Doppler shift) because the car is moving fast relative to your position. You
would not hear this difference if the driver were pushing the race car past you. The
Doppler effect depends on how fast the sound source is moving relative to the speed of
sound. Because sound travels five times faster in water than in air, whales would have to
swim insanely fast to generate any perceptible Doppler shift.

The acoustic details within humpback whale songs provide some clues as to how singers might benefit from enhanced pitch resolution capacities. The most obvious feature of any humpback whale song is its repeating patterns. You can easily hear that the units within those patterns are different. One thing that a human listener can't do perceptually, however, is mentally merge units over time so that units produced sequentially are heard simultaneously. Why would you want to lump consecutive units together? Because this is the kind of perceptual integration that would enable you to use reverberation to judge the distance of a singer if you were a chickadee or a whale.

If you could mentally merge humpback whale song units over time, you would discover that consecutive units within song phrases create a kind of chord, like the combos of sounds that piano players create when pressing multiple keys at the same time. The pitches or notes within this collapsed whale song chord do not sound like those you would normally hear in human music. But they do seem to follow certain rules about how pitches can go together within the chord. The main rules are (1) avoid producing the same pitches when not repeating a unit, and (2) make highly reverberant elements differ in pitch from less reverberant elements.

By following these simple rules, singing humpback whales create long-lasting reverberant acoustic fields that for the most part do not overlap in pitch. When such acoustic fields reach a whale's ear, each reverberant stream will register at different locations along the whale's cochleae. In this way the reverberation from each phrase within a humpback whale song ends up being something like a barbershop quartet, if you imagine that each "barber" is a different gigantic species of horseshoe bat.[16]

What happens when chord-like streams of reverberation from songs hit the highly innervated ears of humpback whales? Given

16. Different bat species produce echolocation cries at different pitches.

the exceptionally large number of neurons devoted to each segment of a whale's cochleae, it's likely that separate sets of sensory neurons are activated by each reverberant stream. In effect, singers are creating an acoustic web, one that potentially can reveal the presence and movements of any large objects that enter it. A singer's ears are at the center of this web. An exceptional capacity to detect and discriminate different modulations of pitch should increase a singer's ability to identify when specific streams are fluctuating, in much the same way as an echolocating horseshoe bat.

Now the potential advantages of sedately hovering while singing become clear. Much like the perching horseshoe bat, the stationary singer gains a relatively static acoustic background against which to detect any whales entering his acoustic web. The longer a stationary singer (or perching bat) remains in a particular locale, the more time its brain has to adjust to the local soundscape. Neither singing humpback whales nor echolocating horseshoe bats are anchored to a single spot, however. Both bats and whales move from one spot to another, and both species vocalize when not stationary, presumably to maintain their awareness of their surroundings while they're on the move. Singing humpback whales and echolocating horseshoe bats are like nomadic spiders that can pull up their web and bring it along with them to check out multiple hangouts. In this scenario, singers are stationary because being stationary is an echoic hearing aid.

A key difference between echolocating horseshoe bats and singing humpback whales is that the bats monotonously produce the same cries for many hours each night and do so throughout their lives. Singing humpback whales, in contrast, produce a huge variety of units within each song and dynamically shift the properties of units and phrases over time. How does this extreme versatility relate to the listening tasks that singing whales create for themselves and others?

Whale Song Acoustics: Songs as Scenes

Some early analyzers of whale songs attempted to figure out whether singers that were audible to each other might be in some way singing together. They wondered if "choruses" of singing whales might resemble a human chorus—for instance, by synchronizing sound patterns or rhythms, or maybe even harmonizing. In short, scientists wondered whether whales singing "together" (while spaced many miles apart) coordinated their vocal actions. What they found was somewhat convincing evidence that singing whales do not synchronize songs. Instead, singers seem to do their own thing. As a result, most studies of whale song have focused on analyses of songs produced by individual singers, usually those that are not part of a chorus (because that can complicate analyses quite a bit).

Although whales don't sing in synchrony, they may vocally interact in other ways while singing. For instance, some recordings show that a single whale starting to sing in a region where other silent whales are present can trigger the silent whales to start singing. That sort of coordination could be coincidental (maybe that is the time of day when whales normally start singing), it could be reflexive (like when dogs howl in response to a siren), or it could be a process known as stimulus enhancement, wherein one's actions direct the attention of others (imagine a person pointing up at the sky, causing others to look up).

Preliminary evidence also shows that singers will modify what they do in real time based on other singers' actions. They might lengthen or shorten parts of their songs more than is typical or modify their songs in ways that reveal what they are going to sing when. Rather than ignoring every other whale's vocal actions, singers may actively adjust properties of songs in ways that reduce the likelihood that any two singers will step on each other's toes. In other words, singers may actively listen to what is going on around them and weave their songs into existing soundscapes in ways that enable them to avoid interfering with other singers and/or temporarily overlap with a specific singer while ignoring others.

There's currently no way to know for sure whether singers view other audible singers as background noise or as an acoustic tapestry within which all singers can benefit from closely attending to who is singing what when.

Units from different singers overlap in time.

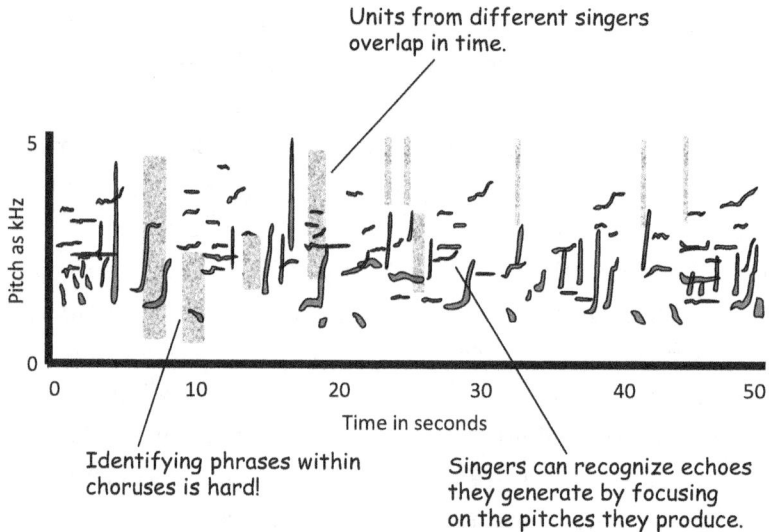

Identifying phrases within choruses is hard!

Singers can recognize echoes they generate by focusing on the pitches they produce.

Listeners must tease apart different vocalizers to localize them and to interpret the sounds they are hearing, despite song overlap within choruses.

Not all singing whales vary their songs as much as humpbacks do. Most singing whales are in step with the horseshoe bats in terms of keeping things simple and repetitive. So it's not the case that singing underwater necessitates changing your song. Presumably, humpback whales gain some advantage from their sonic shifting that is specific to their circumstances. Sexual advertisement hypotheses propose that humpbacks change their songs because novelty is sexy. According to this view, listening females need to hear songs well enough to judge them, but otherwise song structure has nothing to do with listening.

The sonar hypothesis, on the other hand, implies that shifts in songs in some way enhance a singer's ability to detect or recognize echoes. This could work in many ways. Which of these scenarios is actually happening is an open question. One way that changing units could enhance echo reception is by accommodating the environments potentially surrounding a singer. Humpback whales sing in quite shallow water over flat bottoms, as well as in deep water and near rapidly sloping banks, all of which have different propagation characteristics. Singing humpback whales face a much broader variety of sound channels than do horseshoe bats. Perhaps cycling through a diverse set of units increases the odds that at least some usable echoes will make it back to the singer regardless of the environmental conditions.

Along similar lines, producing phrases in a predictable order could enable singers to shuffle through different sets of units in ways that expand their search range. For instance, some phrases might be more effective for generating echoes at the fringes of a singer's detection range, while others might be more effective for revealing targets in shallower regions closer to the shoreline. No set of sounds is optimal for long-distance transmission in all shallow-water conditions, so perhaps scanning for targets at different distances and in various directions can increase the odds of eventually detecting them.

Another issue that singing humpbacks face, but not perching horseshoe bats, are the choruses produced when multiple whales sing off the same coastline. If all singing humpback whales were producing a single unit over and over like horseshoe bats, it would be extremely difficult for each singer to tease apart which echoes are the result of its own units as opposed to echoes produced by other whales' songs. Using different types of sounds with varying pitches may facilitate the sorting of echoes by listening singers. Singers may still hear units and echoes generated by other singers, but as long as they're not producing the same pitches at the same time, each singer should be able to focus on echoes of its own songs. A singer could potentially accomplish this by focusing its attention on the echoic pitches it's producing and ignoring pitches from other singers. The higher acuity provided by having additional sensory neurons could, in this case, enhance a singing humpback whale's ability to selectively listen and attend to the echoes generated by its own phrases.

Ultimately, echolocation is all about hearing the consequences of one's own actions. When multiple individuals echolocate within earshot of each other, however, then discriminating one's personal consequences from those of other echolocators becomes critical. The listening task faced by a singer attempting to echolocate is in most respects similar to that of listeners eavesdropping on multiple singers. If a female or male listener wants to judge the quality of a singer when multiple singers are present, it must somehow separate the songs of each singer and somehow account for the echoes and reverberation generated by those songs. The main difference for the singer is that it knows in advance exactly what sounds it's producing when and so has a preexisting template for what to expect in terms of echoes. Additionally, a singer listening for its own echo streams need not attend to the songs of other singers since the main echoes of interest are those that it generates. The singer's perceptual task is to recognize itself from among others. The task faced by potentially judgmental listeners is to evaluate all audible singers, which requires separating all the songs without prior knowledge of what they were like when they were produced.

Neither listening task is trivial, and scientists are still working out the auditory computations that make it possible for any animal to segregate concurrent sound streams in this way, a topic that will reappear in chapter 9. In the case of echolocation, extracting and interpreting self-generated echo streams is at the crux of the phenomenon. Echolocation is more an issue of reception and perception than one of production, and the presence or absence of ultrasonic clicks is not particularly relevant when evaluating whether whale song functions as a sonar signal.

Historically, the best evidence for animal echolocation has come from behavioral observations and experiments. This is how biosonar was experimentally established in bats, dolphins, birds, and humans. Knowing exactly what a whale perceives when it hears echoes is a challenge that may never be met, but it is possible to monitor and manipulate a whale's behavior. Only one experiment has been conducted to see if humpback whales can echolocate. This experiment involved tethering a humpback whale to the shore and seeing whether it would avoid swimming into pipes placed in front of it when the whale was blindfolded.[17] This is a fine if somewhat ethically questionable test for echolocation in smaller toothed whales but reveals nothing about how baleen whales might use songs as sonar since the signals that best reveal large targets from long distances are radically different from those that might pinpoint the locations of small, nearby pipes. Chapter 8 will consider whether any of the available observational and experimental findings from studies of singing humpback whales provides evidence for or against the idea that singers use self-generated echoes to monitor other whales' distant actions.

_____

17. Spoiler alert—the whale ran into the pipes.

CHAPTER 8

For Whom the Whales Toll

IN THE EPIC *MOBY DICK*, Herman Melville (through the narrator, Ishmael) proclaims that the sperm whale "has no voice; unless you insult him by saying, that when he so strangely rumbles, he talks through his nose."[1] This common belief in sperm whales' muteness was not overturned until the 1950s, when researchers recorded an "overlapping clatter of click series"[2] emanating from a group of five sperm whales. Even in the earliest reports, observers noted the similarities between sperm whale clicks and the clicks that smaller toothed whales use for echolocation. Today, cetologists universally accept that sperm whales use clicks for both communication and echolocation. And yet there's no conclusive evidence that sperm whales actually echolocate. No one has tested a sperm whale's echoic perception with the methods that established biosonar in bats and dolphins. No one knows precisely what sounds sperm whales can hear or how accurately they're able to localize echoes. Why then have cetologists become convinced that sperm whales are echolocators? In addition to the similarities between

---

1. *Moby Dick*, chapter 85; see https://www.gutenberg.org/files/2701/2701-h/2701-h .htm.
2. Watkins (1980).

sperm whale clicks and those used by echolocating dolphins, it's the behavioral context that matters: Sperm whales produce click trains while foraging. In other words, most scientists believe sperm whales echolocate because their behavior in the wild is what one would expect from an echolocating cetacean.

The Nobel Prize–winning ethologist[3] Niko Tinbergen suggested that any animal behavior could be explained with both proximal explanations (meaning in the here and now) and ultimate explanations (reflecting a species' natural history). Proximally, sperm whales click because their brains compel them to click, and their bodies developed in ways that make it possible. The ultimate answers to the question of why they click are that natural selection shaped sperm whales (genetically) to produce clicks, and sperm whales' clicking ancestors fared better when foraging than non-clicking whales.

These same kinds of explanations for "Why?" also apply to singing baleen whales. Whales sing because of how their brains and bodies developed, because their ancestors benefited from doing something like singing, and because singing enhances their ability to survive and reproduce. Note that these Tinbergian explanations apply equally well whether songs are a sexual display or a form of sonar. The sonar hypothesis provides an additional kind of answer for why whales sing that is more psychological than biological—namely, that whales sing to apprehend and thereby to give themselves choices.

Unlike clicking sperm whales, singing baleen whales typically aren't attempting to eat.[4] Consequently, one would expect singing baleen whales to behave differently from echolocating sperm whales. *Some* behaviors should indicate that singers are perceiving

---

3. Ethologists study animal behavior by observing what animals naturally do, usually in the wild.

4. Although as noted in chapter 4, they have been recorded singing while foraging.

events echoically, though, if that's why they're singing. Prominent whale researchers have argued that whale songs can't possibly be a form of sonar because in their view observations of humpback whale behavior aren't consistent with songs functioning as a kind of sonar. I find this conclusion ironic because observations of humpback whales' actions off the coast of Maui were what ultimately convinced me that humpback whales could use their songs for long-range sonar. Hopefully by the end of this chapter, you'll understand why.

Some whale behaviors suggestive of echolocation are impossible for a human listener to detect without the help of statistical analyses. For instance, if you measure how singers vary the amount of time they spend producing different themes within a song session, it parallels how birds of prey inspect different locations within a visual scene. Neither whale watching nor listening to songs would allow a person to directly perceive this kind of behavioral convergence.

While some of the actions of singers are hidden from casual observers, other behaviors are directly observable. By watching what singers do (and do not do) when singing and by watching how other nonsinging whales behave around singers, one can discover clues about how and why humpbacks use songs. Since a singing whale's active space[5] is more extensive than those of dolphins, bats, or sperm whales, monitoring relevant song-related behaviors requires observing a considerable swath of sea for many hours. Only by considering this bigger picture can scientists fully assess whether singing whales behave like long-range echolocators.

---

5. Recall from chapter 5 that an echolocating animal's active space is the three-dimensional extent of the surrounding environment from which the animal can form echoic percepts, analogous to one's field of view.

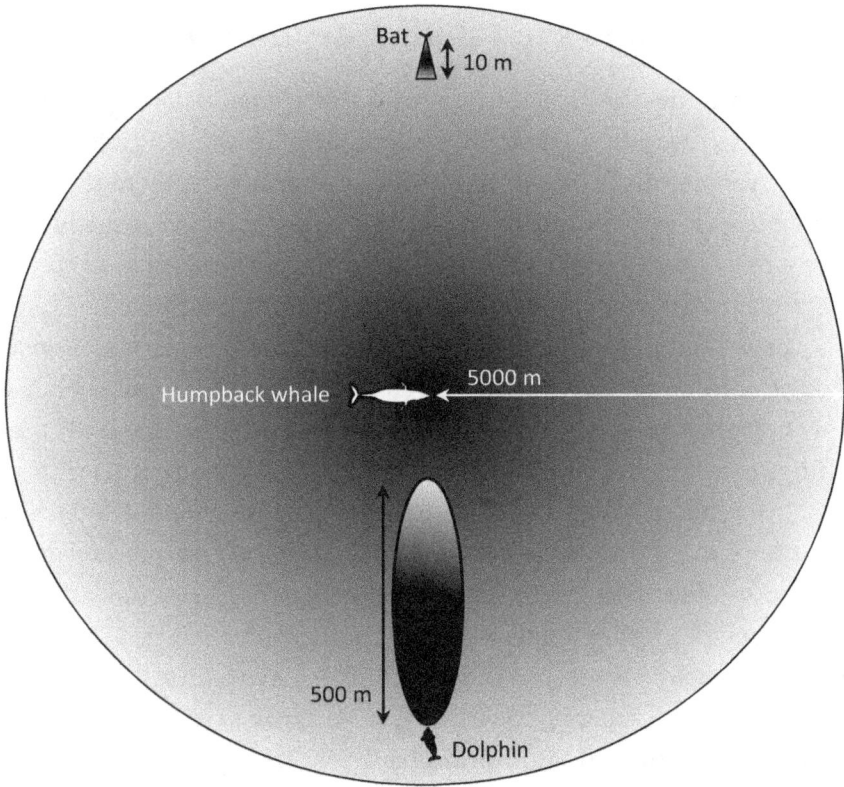

When bats echolocate, they're usually perceiving and reacting to echoes coming from targets located in front of them at distances of less than 10 meters. Echolocating dolphins can detect targets located 50 times farther than this, but dolphins are still limited to perceiving targets in the direction they are facing. Because the sounds within humpback whale songs are lower in pitch than those used by either bats or dolphins, their potential active space is more than 500 times more extensive than that of any echolocating bat and extends in all directions, not just in the direction they're facing. In this respect, the relevant search space of a singing whale is more comparable to that of a raptor flying high overhead than to the active space of a bat or dolphin.

Using Songs to Coordinate Connections

Understanding the interactions of baleen whales is easier said than done. Despite their massive size, they are challenging to observe. Whales are hard to see from boats because they're usually underwater, spaced out over vast distances, and constantly on the move. Shore-based scans help with the spacing issue to some extent but at the cost of reduced resolution. In locations far from shores where water is rough, behavioral observations are almost impossible, and no one has bothered trying to watch whale behavior at night.[6] For these reasons, scientists don't yet know all the things that whales do while singing or all the ways that listening whales react to singers. The surprise discovery mentioned in chapter 4 that humpbacks sing while foraging is just one example of the hidden lives that singers lead. Despite these limitations, sightings of singing whales (mainly humpbacks) aren't rare. Researchers have monitored singers' actions, as well as the actions of surrounding whales, for many years. These past observations provide a window into how humpback whales interact in contexts where singing is prevalent.

Most notably, observations made from both boats and shore stations show that humpback whales regularly sing when no other whales are nearby. More than 80% of singers identified in past studies were alone, often for many hours. The solitude of singers is somewhat ambiguous because other whales could be located many kilometers away, out of sight of any human observers, listening to the singer—a covert audience that some whale researchers have described as a *heard* of whales.

---

6. Researchers can use recordings made from multiple locations to triangulate the positions of singing whales at night, but they can't really tell what's happening other than knowing when singers are stationary versus swimming.

Whether such hypothetical heards exist is unknown. Although it's intriguing to think of whales located many kilometers away from each other as a social group, there is currently no evidence that individuals within such "groups" coordinate their actions in any way. The fact that other whales can hear a singer off in the distance no more implies that the listeners are socially interacting with the singer than the fact that you can listen to Taylor Swift on Spotify implies you are part of her social circle. Listening whales that do not approach or vocally react to an audible singer are eavesdroppers. Scientists are free to believe that lone whales sing to hidden heards, but it's simpler to assume that if you see lots of whales singing by themselves, maybe it's because those singers are alone.

Some cetaceans do show strong social interactions within stable groups of individuals. For example, pods of orcas often consist of family members that spend their whole lives cooperatively hunting and traveling together. Baleen whales are not this kind of cetacean. They're much more likely to mind their own business. There are some exceptions to this general rule. Certain groups of humpback whales coordinate their actions while corralling fish with bubbles, a mind-boggling feat called bubblenet feeding. Individuals within such groups are often seen together across multiple years. But that's where the similarities to other cetacean social groups end. When these humpbacks are not bubblenet feeding, they don't hang out together. And strangers sometimes join the bubblenetters, so they're not an exclusive social group. Whales coordinating their actions while bubblenet feeding provide a model for what socially interacting baleen whales look like: multiple whales swimming in predictable, coordinated formations and performing joint actions with a clear goal.

Studies of bubblenet feeding have the advantage that these cooperative enterprises culminate in spectacular surfacings that researchers can document from boats or using drones. In contrast,

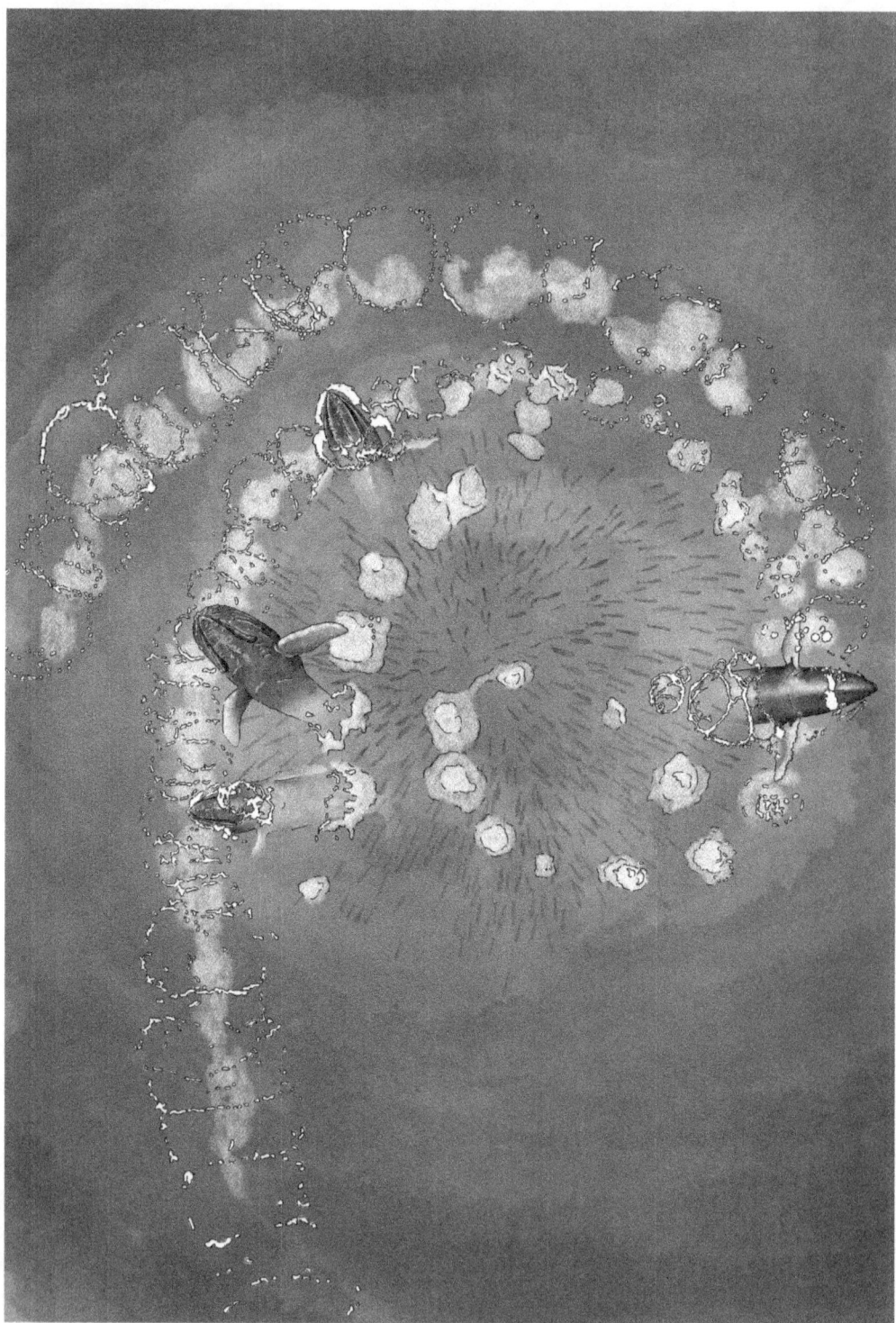

many observations of singing whales involve triangulating their locations by recording them from an array of regularly spaced hydrophones, in which case there's no way to know if the singer is alone or part of a heard. A sample of about a hundred singing humpbacks tracked off the coasts of Kauai using a large array revealed four main ways in which singers moved while singing. Some humpbacks sang while traveling in a straight line but only when they were arriving in Hawai'i or leaving to head back north. The most common behavior (observed in more than half of singers) was singing while repeatedly diving and slowly drifting—this is the prototypical mode of singing. Some singers also *milled* while singing, meaning they stayed at one spot and sang there for a relatively short time. Others did a combo of drifting, milling, and swimming, sometimes with sharp changes in the direction of movement. The likelihood of seeing each of these patterns varied with the time of day. Tracks of singers collected off the coast of Maui showed that singers shifted closer to the shore during the night and then went offshore during the day. Collectively, it appears that singers tracked off Hawai'i changed their spacing over time in relation to one another. When singers shift their locations in this way, it could change their proximity to nonsinging whales. Or it could be that all humpbacks collectively change where they hang out at night regardless of whether they're singing. One of the limitations of using arrays to observe whales is being able to track only a subset of any animals that are present.

Although singing humpbacks are relatively stationary, singers do not stick around at a particular location for more than a day or two. This suggests that whales don't seem to use songs to stake out

Opposite page: *Cooperative bubblenetting.*
Groups of humpback whales will sometimes join forces and encircle a school of fish with bubbles (which the fish avoid) before rising in coordinated lunges to swallow the panicking prey.

individual ocean territories. Some researchers believe, however, that aggregations of humpback whales may use songs to establish a sort of mating arena. Lou Herman proposed in the early 1980s that groups of singing males might be more attractive to migrating females than a single singer. He suggested that communal singing created a *lek*, which is an assembly of males that females visit to select mates. Lou's basic idea was that more singing indicates more males, meaning that females who approach a lek of singers have more potential mates to choose from.

Many different species of birds and mammals form leks as a reproductive strategy. Some key differences are found between terrestrial leks and what humpback whales do on their breeding grounds, however. First, males in terrestrial leks densely congregate at one spot and stay there. Second, individual males within the lek establish their dominance early on. Dominant males are heavily favored by females—other males are effectively vocal spectators. Third, pregnant females generally do not visit terrestrial leks since they're not seeking out mates. And finally, terrestrial lekking males typically have much bigger voice boxes than females, and they produce simple, repetitive advertisement calls.

Cetologists generally put aside these discrepancies. They explain the behavioral and physiological mismatches by noting that marine mammals face different circumstances than terrestrial mammals. It would be relatively straightforward to test the lekking hypothesis in humpback whales—one could continuously broadcast high-amplitude recordings of choruses of singing whales at some location that female humpbacks are not known to frequent. If females are attracted by these mega-concerts, then they're attentive to singer concentrations. If they're not, then game over. Unfortunately, no such experiment is likely since scientists do not want to risk disrupting the natural actions of breeding whales.

It would probably be a waste of time and money to put on such a show, anyway, because female humpbacks are constantly on the move when in and around the breeding grounds. They show no tendency to congregate around groups of singers. Partly that's because there are no "groups" of singers comparable to what's seen in terrestrial leks (aside from those hypothetical heards). Only simultaneously audible isolated singers are spread out over tens to hundreds of square kilometers. Some scientists argue that male humpback whales use songs to space themselves out, dividing up areas where females might show up into temporary territories. Singers are commonly spaced a kilometer or more away from each other, consistent with this possibility.

The lekking hypothesis predicts that humpback whale songs attract females. The spacing hypothesis predicts that these same songs repel other males. Given these predictions, you might be surprised to learn that when researchers have broadcast recordings of songs to whales within breeding areas,[7] the most common outcomes are (1) nothing happens, and (2) males are attracted to the location of the speaker. This second outcome is not due to males being particularly curious about how underwater speakers work. When researchers camped their boats out over singers, they similarly found that the whales that approached were exclusively males.

At this point you might think that the approaching males are encroachers attempting to assert their dominance and usurp the singer from his temporary territory. In this case you might expect there to be aggression, threats, and other such mayhem once the approaching male reaches the singer. What usually happens, however, is that the singer stops singing, the two whales hang out relatively close together for 5–10 minutes, and then the joiner swims away. Observations made by whale researcher Jim Darling revealed that this was the scenario 9 out of 10 times when a male joined a

---

7. These kind of playback experiments are designed so that only a few whales hear the broadcasts and only for a relatively short time, so they are much less likely to significantly affect whales' reproduction than the hypothetical experiment mentioned above in which loud recordings of chorusing whales are continuously being broadcast.

singer. Some potentially aggressive interactions occurred, but just as often the two males swam off together. In a few cases, the singer kept on singing like nothing was happening. Interestingly, joiners hung around nearly 4 times longer when the singer "ignored" them.

Males approaching singers are not just individuals checking out the local scene, because singers and joiners often interchange roles, as in this example described by Jim: "This male, singer B, over a four-hour period, interacted with three, possibly four, males, for five–six minutes each over a distance of thirteen kilometers (km). The singer was joined, stopped singing, then itself joined and split from two separate singers (D and E), then began singing again— and likely was joined again, although this was not confirmed."

In these scenarios it's clear that the joining male does not need to echolocate to find the singer; he can simply home in on the song's origin. But what about situations in which the joined whale is not singing? Consider this example, again provided by Jim: "Two neighbouring singers (2A and E) stopped singing and simultaneously joined a passing cow/calf and escort." And this one: "Travelling singer C stopped singing and joined singer B for less than 3 minutes before split-up. Ex-singer C then raced 6.5 km (7.5 km per hour) until it joined a cow/calf and escort."[8]

A quick note here on the social groups typically found in the breeding grounds: They are lone females, single females with their calves (the cow/calf), cow/calf pairs accompanied by another adult whale (called an escort), cow/calf pairs accompanied by multiple escorts, all-male groups of various numbers, females accompanied by one or more males, and lone males. These social groups constantly fluctuate, with individuals potentially participating in all the various combinations during a single week.

---

8. Quotations from Darling and colleagues' 2006 paper, "Humpback whale songs: Do they organize males during the breeding season?," pages 1064 and 1066.

In the examples above, the singers stopped singing just before swimming to join a distant group of whales. The singers knew the likely location of that distant group prior to departing to join them. How did the singers know where that group was heading? Visibility underwater is low over kilometer ranges. Most whale researchers agree that the singers must have used their sense of hearing to localize the distant whales. But for that to happen, the groups of whales that were joined must have produced sounds, which the singers must have detected despite simultaneously producing their own intense songs. Jim Darling makes no mention, however, of the joined pods producing sounds (by vocalizing or slapping body parts on the surface of the water) prior to being approached by the singers. If those groups were not broadcasting their whereabouts when the singers detected and localized them, then the singers could have determined the groups' presence or headings in only one way.

As far as I know, no scientist has ever managed to localize and track a cow/calf with escort from a kilometer or more away using any underwater technology, except in cases where the escort is singing, a scenario I'll consider in more detail below. Researchers typically locate and monitor such groups by watching for blows when the whales surface. This is not an option for singers. Such social groups do not continuously broadcast their locations like ice cream trucks and for good reasons. The escort, almost always a male, would not benefit from alerting other males to the fact that he's with a female, and there's a potentially high cost to mothers if large males physically compete to be close to her with a calf nearby.

Cows with and without calves employ several strategies to avoid males, including spending most of their time in places males rarely go, being silent or quiet, and swimming away from locations where singers are present. They may even engage in evasive maneuvers to avoid being joined by a singer, per this report by whale icon Peter Tyack:

While a singing whale I was followed for 7 h, there was only one obvious interaction between I and another whale, a nonsinging lone adult, E. E was first sighted at 1411[9] during a period when whale I was moving very slowly, 100 m in 19 min. E on the other hand was moving rapidly (3.5 km/h). . . . At its 1425 surfacing, E had moved to within 400 m of the 1418 surfacing of I. Singer I then accelerated to match E's velocity. By the next surfacing of I at 1433 and of E at 1436, the gap between the two had closed to 260 m. Both whales then made a right-angle turn after which whale I surfaced even closer to E, but E then turned back to its original course. The singer, I, did not follow at this point but slowed down dramatically.

In this case, the singer was clearly trying to join a passing whale but failed to make it happen. Given that the passing whale performed evasive maneuvers exactly when the singer was closest to joining, at the time when the singer began to swim more rapidly, it is unlikely that the nonsinger was interested in being with the singer. How then did the singer know that "whale E" was even in the area? Again, the researchers observing this interaction make no mention of whale E producing any sounds.

In this example, why did singer I seemingly only detect and approach the nonsinger when it was closer than 300 meters away? Relatedly, why was nonsinger E swimming in singer I's direction if it was not interested in interacting? The answer to both questions may be the same. As noted in chapter 7, the strongest echo reflecting from a whale will come from its lungs, which means that the weakest echoes will be returned when a whale is swimming directly toward or away from a singer. The head-on or tail-on position reflects the weakest echoes because this exposes the smallest lung surface area. In contrast, a broadside position reflects the

9. 1411 corresponds to 2:11 p.m.

strongest echo. Notably, in the example above the singer made its attempt to intercept just when the nonsinger was swimming past (when the broadside of the nonsinger's lungs were exposed to the singer), and this was also when whale E made a 90-degree turn. Such a turn would convert a maximal- strength echo into a minimal-strength echo. In other words, by swimming toward the singer and then performing two 90-degree turns, the nonsinger potentially minimized the song-generated echoes from its body, possibly as a detection-avoidance tactic.

Throughout the scientific literature are many observations of singing humpback whales joining distant nonsinging whales. It's common in such cases for researchers to assume that females or other whales are either vocally soliciting singers or inadvertently advertising their movements. It's true that nonsingers do make sounds, both vocalizations and percussive sounds, especially during social interactions within larger groups. And singers are known to approach groups making sounds, as well as broadcasts of recorded social sounds. This does not imply, however, that every instance of singers approaching distant whales depends on those whales making sounds. Examples such as the one above suggest that at least some whales, including cows with calves, are motivated to keep their locations or movements acoustically hidden.

Whitlow Au and colleagues claimed that observations of male humpback whales approaching singers were inconsistent with songs serving a sonar function. But both dolphins and bats commonly approach areas where other individuals are echolocating. The fact that other animals can hear sounds and make use of that information is in no way evidence against those sounds functioning as sonar. When bats home in on other echolocating bats, it's probably because they're looking for food. Why are male humpback whales homing in on singers? Jim Darling suggested that songs may provide a way for males to form social relationships, including potential alliances that could prove useful in future physical competitions. Other researchers have proposed that male humpbacks approach singers to see if there's a female with the singer.

Past observational studies found that 10–20% of singers hung around with a cow/calf at some point during a song session. Consequently, a male that swims from singer to singer might eventually encounter a cow/calf. Females with calves are not the best option for males hoping to reproduce because mothers are often less receptive and fertile when caring for a calf. There are documented cases of mothers becoming pregnant, however, and some chance at mating is better than none. It's unlikely that males singing in the presence of any female are inviting other males to come compete with them, leading some researchers to propose that a male singing with a cow/calf is attempting to woo the mother. In this scenario the fact that the singer is inadvertently attracting the attention of other hopeful bachelors is viewed as an unfortunate side effect.

How likely is it that a singer in the company of a cow/calf is serenading the mother? One way to assess this is by looking at similar situations in other species. I mentioned back in chapter 2 that there are parallels between the mating behaviors of humpback whales and mountain sheep. Recall that mountain sheep are genetically close cousins of baleen whales that mate during a rut (a breeding season). Rams don't sing to nursing ewes, but their actions during the rut can potentially shed some light on what's happening when an escort sings with a cow/calf nearby.

Rams use multiple tactics to locate and gain access to receptive females. If one ram starts showing interest in a female, others will hurry over to investigate. Groups of rams will gather behind a female and follow her when she's about to be in heat. When a female comes into heat, the largest ram stays close to her and attempts to keep other following males away. Intense physical competitions take place between males to establish which one gets to be closest to a female, some of which last hours. Losers are not chased away;

subordinates will either hang around or leave. A third ram often interferes in such fights, and as many as six may interfere. In some cases, rams form temporary alliances in these competitions. If a female attempts to flee while the males are fighting, they may chase her. If she's not yet in heat, she'll perform evasive maneuvers to escape. A winning ram hangs around the female without physically contacting her until she finally gives in. Rams are with females a relatively short time, leaving soon after mating; the actual reproductive act is quite quick. This nonterritorial mating system is called a *tending bond*.

Now read that last paragraph again and replace the word "rams" with "male humpback whales." Aside from the fact that it's much trickier to tell when a female whale is in heat, the behavioral scene is the same. Male humpbacks seek out females, then appear to "guard" them. They often follow a single female and compete to be the closest. The whale that's closest to the female attempts to repel other males. The competitions are fierce and prolonged, with males often drawing blood (in at least one documented case, a battling male humpback whale died). In some cases, pairs of male humpbacks seem to coordinate their attacks. The female within these groups often swims at the head of them and sometimes seems to attempt to get away from the group. Generally, these social groups are short-lived, lasting no more than a day. Surprisingly, mating has yet to be documented in humpback whales, which suggests the grand finale is either quick or occurs mainly at night.

In mountain sheep the ruggedness of the environment contributes to the mating tactics of rams. The mountainous terrain makes it easier for younger rams to evade the dominant ram and mate with females. Younger rams use a few strategies for this. Sometimes a young ram will guard a female not in heat. He may block her from approaching areas where dominant males would find her. Smaller rams may also provoke aggression in a tending ram, causing the ewe to flee. This gives the smaller rams an opportunity to chase and mate with her. These strategies are not as successful as tending, but they do work! Tending bond mating systems are prevalent in mammals that live in places where it's difficult to defend resources, especially when the density of receptive females is low, and the environment makes it harder for an individual to effectively herd groups of females. All these factors are present in the breeding grounds favored by humpback whales.

From the available behavioral evidence, a female humpback's choice of mate seems to be somewhat limited. The winner of the Oceanic Gladiator competition is probably the male that gets a chance to mate. In such macho chase scenes, the main ways a female can "choose" are by either rejecting the winner's advances or soliciting the attention of potential challengers in the area. Nevertheless, many whale researchers assume that female choice is a dominant force in the humpback whale mating system. Why? As Lou Herman put it, "It is difficult to contemplate the complexity, diversity, and richness of song as not being driven, at least in part, by female choice."[10] This difficulty has befuddled generations of cetologists. Today the mating system of humpback whales is typically described as some combination of tending and nonterritorial lekking based on vocal displays by singers that rarely attract females but may foster alliances between males or help males space themselves apart.

This brings us back to those singing escorts accompanying cow/calves. Here's a situation, some say, in which the female may contemplate the virtues of the serenader's songs up close and personal. It's also a scenario that some critics claim makes the sonar hypothesis implausible. Whit Au and colleagues argued that the presence of singing escorts "nullifies" the sonar model because if males use songs as sonar, then they should not sing while accompanying a female. The thinking here is that if males use their

---

10. From his 2017 paper on the multiple functions of male song, page 8.

songs to search for mates, then once a female is found the search is over. One need only head to a local bar to discover the flaw in this logic. Male humpback whales are not simply searching for females; they are searching for females they can impregnate. If a male could continue to survey the social scene at no cost while hanging out with a nursing mom,[11] then why wouldn't he?

In summary, singing humpback whales interact with each other and with other nonsinging males and females in lots of different ways. The structure and composition of groups of whales on the breeding grounds is constantly in flux: Silent males may approach and join singers, singers sometimes approach other whales and sometimes don't, and sometimes whales sing while they are accompanying another whale. It's certain that songs affect how whales socially interact. But the role that songs play in those interactions is nonobvious. The sonar hypothesis does not attempt to explain the social actions of whales in any detail because it is fundamentally a perceptual hypothesis. Knowing that humans can see is not going to give you many insights about how teens will socially interact during a party, and similarly, knowing that singing whales can echoically perceive events from long distances is not going to tell you much about what a party of whales is likely to do.

Proponents of the sexual advertisement hypothesis claim that females listen to the songs of potential mates and then express their preferences for the singers that produce the most novel and/or complex songs, with the caveat that all singers in a particular region sing the same song. They presume that female whales are like oceanic versions of peahens, judging males' acoustic tail trains. How females do this, no one knows. The properties of a singer's song that reveal his prowess and turn female whales on are a mystery. In peacocks, females prefer high numbers of eyespots, which are

11. Who might or might not go into heat at some point.

directly related to a male's size. Whale researchers who believe that songs are reproductive displays are heavily invested in the idea that females prefer some songs over others and that the main goal of singing whales, especially humpback whales, is to produce songs that are maximally alluring. Given that the songs of humpbacks are constantly changing, this view on song function implies that sexiness requires state-of the-art complexity and the near-constant learning of new song variants.

Social Learning and Vocal Culture in Singers

IN THE SOUTH PACIFIC—A HUMPBACK WHALE KARAOKE LOUNGE; WHALES PICK UP A CATCHY TUNE WHEN THEY HEAR IT; A HUMP-BACK WOODSTOCK; HUMPBACK WHALES PASS THEIR SONGS ACROSS OCEANS: These are headlines from *The New York Times* on the vocal worlds of singing whales. Reports of whale song fashions fascinate the public, probably because of what they imply about the minds and lives of whales. Such discoveries also intrigue many animal researchers and cognitive scientists because so few species show a comparable capacity to learn from their peers.

For a long time, researchers believed that only humans learned from social tutors. Like language, this capacity was thought to be one of those unique features that only "intellectually superior" humans could manage. Today many examples of animals learning new skills from other animals are known. When it comes to singing, most of those examples involve birds learning which songs to sing by hearing and watching other birds sing. When birds learn songs, it's so they can later use those songs during social interactions with potential mates and competitors. If for some reason a bird fails to learn the songs its peers are singing, that bird's chances of finding and keeping a partner decrease. It's no surprise then that when scientists discovered that humpback whales constantly change their songs, they immediately interpreted this as further evidence that baleen whales use songs as reproductive advertisements.

Researchers realized early that songs changed across the years. But they couldn't be sure if those changes were evidence of learning or just new whales joining the party until they'd recorded individuals across multiple years. Eventually, it became clear that (1) most songs recorded from all individuals in any given year and location were highly similar, (2) songs were never identical from one year to the next, and (3) individual humpback whales were changing their songs from month to month and year to year in ways that maintained their conformity to the population-level norm. Because individual singers within a population were collectively changing their songs while maintaining conformity in song structure, researchers became convinced that humpback whales learn songs from each other through a process known as cultural transmission.

Cultural transmission occurs when individuals learn a particular way of performing some behavior from interactions with others. The learned actions are often described as traditions or customs. In the case of humpback whales, the songs used in any given year are viewed by many as vocal customs. In human cultures, children and adults are introduced to new songs either by hearing other performers or by composing their own songs. You may hear hundreds of new songs each year while learning to reproduce few or none of them. Most of the songs that humpbacks hear each year, however, are basically the same because all the whales within a population are singing highly similar songs. And if the whale hearing the song is a singer, then much of what it hears in any given year will be similar to the songs it is already singing. For songs to change at all in this scenario, singers will have to do something new.

One of the surprising features of the communal, continual change in humpback whale songs is the degree of change, which varies considerably from one year to the next. Some years, singers make just a few tweaks, while in others they totally revamp the shared song. When researchers attempted to closely monitor how whales changed songs in Hawai'i, they found that singers introduced changes relatively rapidly during the breeding season and at different rates as the season progressed. In contrast, songs changed very little during the times whales were not in Hawai'i.[12] What determines how rapidly songs change within a population of humpbacks isn't known, but some researchers speculate that the density of singers—in other words, how often singers hear each other—may be a factor.

Modern studies of humpback whale singing focus heavily on understanding where and how songs change. For instance, recordings made near Australia revealed that humpbacks singing along the east coast adopted a song variant that had previously been used by singers on the west coast. The populations of humpback whales on either side of Australia are genetically distinct. They don't appear to socially interact during breeding seasons, so researchers were surprised to see this kind of coast-to-coast cultural transmission. Later work showed that this same song continued to pop up in other nearby populations of whales over a period of about five years, leading the investigators to describe the phenomenon as a cultural wave.

How and why did this popular song spread across populations of humpback whales? The prevailing explanation is that singers along the eastern coast of Australia at some point heard the songs of their western neighbors, memorized those songs, and then copied them. Other populations further to the east then presumably heard those whales singing the song and likewise copied it, and so on, in a kind of oceanic telephone game.

Cross-whale copying potentially answers how the song spread across vast distances, but not why. For an explanation, whale

---

12. When they were migrating and foraging.

researchers turned to evolutionary theory and logic. If singers are glomming onto the songs of their neighbors, then doing so must improve their evolutionary fitness in some way. And if whale songs function as reproductive displays,[13] then copying the novel songs must somehow increase their effectiveness. Ergo, novel songs must be more attractive to receptive females and/or more discouraging to other males vying to be selected by females. From this perspective the everchanging nature of humpback whale songs is driven by the predilections of potential partners, and the complex features of song structure are little more than a fashion statement.

A key assumption underlying this portrayal of vocal culture in singing whales is that humpback whales are capable of memorizing and reproducing songs and song elements. Singers can only appropriate novel songs or song features if they can recall what they've heard other whales singing and then reproduce it. In other words, for singers to be able to learn to use new songs or song variants they've never heard before, they must be able to vocally imitate the sounds and sound patterns within songs.

Vocal imitation is a rare ability in mammals. Plenty of birds, like parrots, mynah birds, and the superb lyrebird, are adept at imitating all kinds of novel sounds. In contrast, the terrestrial mammals that can vocally imitate are humans and . . . that's basically it. Most mammals that can vocally imitate sounds are marine mammals. Scientists who worked with belugas and bottlenose dolphins serendipitously discovered their imitative skills. Neuroscientist John Lilly, whom you met in chapter 2, was the first person to record a dolphin imitating human speech, a discovery that changed him forever. John's observations were so unexpected that many scientists viewed him as delusional. Decades later, another neuroscientist,

---

13. Like bird songs.

Sam Ridgway, had a similar surprise when he heard a beluga imitating human speech.

Dolphin and beluga "speaking" attempts are not as impressive as those of birds in that they sound less like a person producing words. But what they lack in finesse they make up for in flexibility. Imitative birds typically have to hear a sound many times before they will spontaneously imitate it. Dolphins can make a go at imitating any sound after hearing it once and can learn to imitate novel sounds on command. It may take them several attempts before they manage a reasonable approximation, but they show more attention to detail in reproducing the number, rhythm, pitch, and timing of imitations than do most birds.

The ability of toothed whales to vocally imitate novel sounds provides a proof of principle that at least some cetaceans have the neural hardware required to copy sounds and sound patterns. Their abilities do not imply, however, that all cetaceans can do this—recall that humans are the only vocally imitating primate. Presumably, the factors that favored the evolution of imitative abilities in toothed whales might also have contributed to baleen whale evolution. But what were those factors and how relevant are they to the lives of different species?

Scientists have attempted to answer these questions in bottlenose dolphins by closely monitoring the whistles they produce at different stages of their lives and in different social situations. More is known about vocal imitation in bottlenose dolphins than in any other cetacean because of the relative ease with which they can be maintained in captivity and their tendency to hang out in coastal habitats. In the wild, dolphins are thought to use vocal imitation as a way of learning to produce the kinds of whistles adults around them use, much like children learn the languages used by their family. Dolphin vocal learning differs from song learning by birds because dolphins of all ages whistle in a variety of social contexts.

In the last few decades, scientists have focused on a specific class of whistles that dolphins produce when they're isolated. Soon after public interest in captive dolphins surged in the 1960s, researchers noticed that dolphins repeatedly whistled when put in a tank alone and that bystanders could recognize which specific dolphin was protesting. Human listeners were able to do this because each isolated dolphin produced a distinctive whistle. Researchers dubbed these *signature whistles* to capture the idea that they were as individually specific as a human's written signature. In fact, dolphin researchers hypothesized that dolphins used these whistles as vocal signets to identify themselves to other dolphins. Later field studies revealed that dolphins also used signature whistles during social interactions, especially in situations when they were relatively far from other dolphins and could not see them.

Signature whistles are not the only whistles dolphins make, but they are the main ones studied in relation to vocal imitation. That's because researchers discovered that dolphins that know each other sometimes copy the signature whistles of their comrades. Additionally, some dolphin calves develop signature whistles that are similar to their mother's, suggesting that the young dolphins use their mother's whistle as a model. Male dolphins that travel together in stable groups were also found to modify their whistles to be more similar. Because of associations between vocal imitation and signature whistles, and between signature whistles and social interactions over long distances, most researchers believe that vocal imitation evolved in bottlenose dolphins to facilitate complex social communication, rather than to enable listening dolphins to assess the sexiness of the whistler.

Unlike dolphins, the evidence that humpback whales or any other baleen whales can vocally imitate sounds is sparse. The main phenomenon that researchers point to as evidence of vocal imitation in humpback whales is the dynamic convergence of songs within a population across years. Humpback whales do not swim with family members or social partners like many toothed whales and so have limited opportunities to learn a vocal repertoire from close companions, like dolphins and birds. Anecdotal reports indicate that singing humpback whales sometimes modify unit features to match the sounds they hear while singing. No scientists have attempted to conduct field experiments, however, that would definitively show that baleen whales can flexibly imitate sounds.

The most extensive studies of vocal imitation in any species are in songbirds. Many of these studies focus on the vocal development of stereotyped songs by youngsters imitating their parents and neighbors rather than on adults copying the songs of intermittently encountered strangers. That's because most songbirds do not change their songs much once they are adults. The ones that do are typically adding songs to their repertoire, as opposed to continuously tinkering with the acoustic properties of songs they have mastered. This makes it tricky to apply what's known from studies of vocal imitation in birds to the case of singing humpback whales.

While songbirds that change their songs as adults sing many different songs, singing humpbacks consistently converge on a single song. Some whale researchers have proposed that different phrases within songs might be more comparable to birdsong. Even then, whales use fewer phrases than any songbirds that modify their songs as adults. No songbirds that I'm aware of cycle through their song repertoire in a fixed order like whales cycle through phrases.

Humpback whales don't need to be exactly like songbirds, dolphins, or humans to have a vocal culture. On the other hand, the fact that singing humpback whales communally converge on a song form each year does not, by itself, show that humpback songs are vocal customs that singers continuously learn by copying the latest fads. Nor does it imply that the mechanisms driving whales to change their songs are similar to the song-learning mechanisms in

BLURRING THE LINES BETWEEN VOCAL CULTURES

Humpback whale songs are called "songs" because of birds. If birds didn't sing, neither would whales. Scientists' current claims that whale songs are advertisement displays? Again, birds are to blame. Avian minstrels defending territories and courting cuties set the standard. Similarly, the consensus view that humpback whale songs "culturally evolve" through a process of cultural transmission in which males pirate the songs of their neighbors is, you guessed it, a gift from the world of birds.

Unlike song production and vocal displaying, however, the cultural transmission of songs by birds is limited to a few species. It's especially rare for songbirds to continuously modify their songs after becoming adults—so rare that one can list the bird species that do this on one hand and have a finger or two left over. The caciques mentioned earlier fall into this category. The other songbird commonly compared to humpback whales is the village indigo bird, a small deep-blue bird from Africa.

Like caciques, the reproductive system of the village indigo bird differs radically from that of any whale. These are parasitic birds that lay their eggs in other species' nests, specifically the nests of fire finches. When breeding, males establish a territory from which they chase other males. Females fly from one territory to the next before choosing a mate. Typically, only one male in a local group is mating with most or all of the females.

Male village indigo birds sing around 20 different song types, some of which are copied from other birds like their fire finch caretakers. Birds within a breeding area share most of those song types. All village indigo bird songs last about a second and typically contain from three to nine notes. Indigo birds use some song types to attract females, some to court females, and some when chasing away other males.

Most males in a breeding area sing the same songs from one year to the next. A few change their repertoire if they hear other birds singing different songs, but most of these introduced songs do not catch on. Although the birds keep singing the same songs, they do gradually modify the features of some notes within some song types. It's this gradual morphing of note features across the years that has prompted comparisons with singing humpback whales. These note modifications are subtle, however, and it's doubtful whether anyone who is not a bird expert would be able to detect them. There's also no evidence that female indigo birds pay any attention to these minor tweaks.

If you squint your ears and mind, you might be able to detect some similarities between the vocal antics of village indigo birds and the sound sequences produced by singing humpback whales, but you're going to really have to want it. If any songbirds were vocalizing like baleen whales, it would be an amazing example of evolutionary convergence. As it stands, claims that whales sing like birds are a bit of a red heron.

birds. Sometimes it's reasonable to accept experts' interpretations of what's happening. Other times, expressing an alternative take is warranted and even necessary.

I learned this lesson the first time I participated in field studies of humpback whales in Hawaiʻi. My participation was by decree rather than by choice. Lou Herman required all his graduate students to spend at least a couple of weeks assisting with the fieldwork on humpbacks, both so they would be familiar with the research and so he could take advantage of the additional help.

Fieldwork requires a lot of hours of labor of all types. Coincidentally, the second week I was scheduled to "volunteer," Lou himself decided to join the team out in the field. This was a relatively rare event because usually he was deeply embedded in dolphin research, and the field studies were conducted off the coast of a different island. The second day after his arrival, Lou decided to captain the research boat that went out every morning in search of humpbacks. I happened to be assigned to the boat that day and was intrigued to see how going out with the head honcho would compare to earlier trips captained by grad students.[14]

As it turned out, whales were scarce that day, so Lou decided to give the volunteers on the boat a tour of the coastline. Perhaps I should have been listening attentively as Lou pointed out various landmarks along the shore while narrating their historical significance. But I guess I was still bummed that no whales had shown up that morning. I noticed that some of the waves seemed to be cresting and that off in the distance, a particularly large bulge in the water was rising and heading in our direction. I'd never driven a boat in the ocean, but my intuition from having played on many beaches as a child was that it would not be a good thing if a wave of that size were to crash on top of our boat. But surely, I thought, Lou would not have driven the boat to a spot where there was any chance of something like that happening. Who was I to second-guess his judgment? And did I really want to interrupt his storytelling to reveal my seafaring ignorance? I did not, but I tapped him on the shoulder anyway. Lou looked up at me confused and perturbed, and I pointed past his head in the direction I was looking. He turned and said, "Oh, shit!"

Then Lou swung the wheel around and pulled the throttle all the way down. The transition was so fast that initially I don't think

---

14. This initial stint was during my first year in Hawai'i, when I was still focusing almost exclusively on cognitive studies of dolphins.

any of the volunteers knew what was happening. It didn't take long, however, for everyone to realize we were accelerating toward a quite large and growing wave and that the top of the wave was thinning. Right before we reached the wave, Lou yelled, "Hold on!" And then we were airborne, with the propeller whining wildly as it left the water. Lou beat the wave. Granted, the landing possibly loosened a few passengers' fillings. From that point forward, I no longer feared looking clueless in front of experts. When I see an incoming wave, I point to it. And when I see problems in scientific explanations that others seem not to, I point those out too.

In the case of song copying by singing humpbacks, it took me three decades of analyses before the singers themselves made me see that songs could not be vocal traditions. Until a few years ago, I was convinced that humpback whales converged on similar songs because they copied song variants they heard. I even published a paper titled "Song Copying by Humpback Whales" that described in detail the features of songs that singers shared versus those that varied across singers.[15] Like other researchers, I couldn't come up with any alternative explanation for how humpback whales collectively change their songs year after year. In my mind they simply had to be choosing what to vocally imitate by hearing other whales.

I found some aspects of this song-copying process odd, however. For one, there were some themes within songs that people reported in multiple populations of humpback whales living in different oceans. Why would whales that cannot hear each other use such similar themes? I discovered a few such themes myself. They were not hard to find once I began looking for them. Another oddity was that several of these worldwide themes fell out of favor for several

---

15. This, by the way, was the paper I mentioned in chapter 5 that was initially rejected by the editor but then ultimately accepted after I rejected the editor's rejection. It was published in the journal *Animal Cognition* in 2005.

years only to later reappear in the same population. Like the singers themselves, the themes would emerge, then disappear, emerge again, then disappear. These recurring themes would presumably be familiar to most of the whales in a population so would not have the allure of novelty. Also, the progressive changes in songs were hypothesized to result from either innovations or modest tweaks to unit combinations. It wasn't clear to me how those mechanisms could lead to reappearing themes.

As an example of an even more surprising feature of song change, whales located thousands of miles apart who had no opportunity to hear each other sing appeared to change their songs along the same trajectories. When whales in Hawai'i started to drawl specific units in a theme, whales off the coast of Mexico also began to drag out those same units in their songs. The chances of whales in different populations consistently introducing the same innovations are low. These kinds of parallel song transformations in isolated groups of whales showed that something beyond vocal imitation was determining how singers change their songs.

Over the years, I discovered several transnational song themes shared across populations. No one else analyzing songs seemed interested in these oddities. They rarely mentioned them in their reports and made few attempts to explain them. Then in 2020, I made an accidental discovery that completely erased my prior beliefs about how and why humpback whales change their songs throughout their lives. While analyzing how singers recorded off the coast of Puerto Rico in the 1970s were morphing units, I noticed that those singers produced a theme that matched one recorded off the coast of Colombia in the 2010s. I'd also seen this particular theme in some Hawaiian songs. The shared theme clearly had the same basic structure across these populations isolated in space and time, and I grew curious about how closely its acoustic features matched between populations. When I compared the singers from different regions, I was startled to find that the rhythmic timing and pitch changes of units within this theme matched across the three populations. Themes that isolated groups of whales have separately improvised and adjusted over centuries should not match so closely. To confirm this correspondence was not just a fluke, I began looking at analyses of Hawaiian songs that Jim Darling published,[16] comparing his theme descriptions to the Puerto Rican themes.[17]

As I'd suspected, the singers Jim recorded produced the shared theme in much the same way as the Puerto Rican singers. I began to wonder if the Hawaiian singers might be producing any other themes that were similar, and so I systematically compared each theme recorded in Puerto Rico to all the themes that Darling recorded from Hawaiian whales. The more I compared themes, the more shocked I became. Every theme produced by the Puerto Rican singers was also used by the Hawaiian singers. The phrases were not identical, but the parts that matched contained the same kinds of units in the same patterns with the same timing. And whales from both regions produced the themes in the same order. The Pacific humpbacks and the Puerto Rican singers, separated by two continents and 40-plus years, were singing the same song!

Seeing the same songs used by these separate groups convinced me of two things: (1) humpback whales were not changing their songs as a kind of fashion statement, and (2) innovation and copying errors were not the main drivers of song change. Think about the development of food preparation traditions in different populations. What are the odds that if you sat down to eat in two communities on separate sides of the globe, say a thousand years ago, that both communities would serve you the same meals with

---

16. In his 2019 paper describing songs recorded in the Philippines, Japan, Hawaii and Mexico from 2011 to 2013.

17. Luckily for me, Jim Darling published detailed acoustic representations of the themes he recorded rather than the symbolic descriptions favored by most other researchers.

each of the same foods positioned on the same places on your plate? The likelihood that two populations of whales that were irreversibly modifying their songs every year for thousands of years without hearing each other would end up producing the same patterns in the same order with the same timing is even lower.

Faced with the reality that singing humpbacks were doing something different from what everyone thought, I began reassessing past studies[18] of "cultural evolution" in whale songs to try to reconcile scientists' past conclusions with this new anomaly. The phenomena most often mentioned as incontrovertible evidence of vocal culture in humpback whales are the cultural waves of song change emanating from the coasts of Australia. As noted above, researchers attribute these waves to cross-population cultural transmission, with whales in eastern populations always copying their western neighbors. Again, oddities arise when you examine the phenomenon more closely. Are singers from the far west the only innovative or error-prone whales? If not, then why are the waves only traveling in one direction? Why don't innovations or melodious errors produced by more easterly whales ever catch on?

Those oddities disappear, however, if you assume that none of the populations are copying their neighbors. Recall from the discussion above that singers in isolated populations are known to change their songs along similar trajectories without being able to hear each other. All that needs to happen to create the appearance of a cultural wave in songs is for adjacent populations to initiate song changes sequentially. The factors that determine how rapidly whales transform their songs are not known, but some evidence suggests that singers adjust songs more rapidly when more singers are around. Consequently, the rate at which songs

---

18. Including my own.

Whale Song Acoustics: Songs as Memes

If you've been following the buzz about singing whales in
Australia, you may have heard that singers there have been
part of a cultural revolution (actually, multiple revolutions)
in which large groups of singers in one area appear to have
plagiarized the songs produced by singers of a nearby population. Researchers studying these populations have described
this phenomenon as a kind of cultural reappropriation in
which singers become enthralled with the songs of their
neighbors and, like good fashionistas, immediately begin
copying them. What's so amazing about these Australian hot
hits, you ask? No specific acoustic properties appear to differentiate the "in" songs, so some researchers think it is simply
the newness of the songs that makes them irresistible.

From this perspective, it's not so important that you (or
listening whales) know anything about what any given song
sounds like. What's important is how different what you're
hearing now is from what you've heard before. Like Jack Skellington discovering Christmastown in the movie *The Nightmare Before
Christmas*, you need not understand what you're experiencing to
recognize that it is something completely strange and new. And
this newness may in some cases compel you to try it out for
yourself. New ideas or practices that quickly spread through a
population (think dabbing) are sometimes called memes. For
many memes the details of the actions or symbols involved are
less critical than their distinctive novelty. If singers copy songs
they hear others singing simply because they are impressed with
the newness of the songs, then the songs are effectively memes.

Researchers claiming that singing whales faddishly copy the
songs of neighbors have produced a scientific meme. Scientists
believe singers are copying each other because whales can be
found singing similar songs to their neighbors not too long
after those neighbors were observed singing the songs. This
correlation is viewed as an indication that the use of specific

songs has spread from one population to another. The fact that two similar events occur sequentially, however, does not imply that one event caused the other. Flowers typically bloom in the southern regions of countries before they bloom in the colder northern regions. This is not because northern flowers are copying their southern neighbors or because blooming is a fad that captures the imaginations of the youngest flowers but because the triggers that lead to blooms occur later in the north. We don't know exactly what drives singers to change their songs, so we have no way of knowing if these triggers are occurring sequentially for adjacent populations of whales.

Until scientists know more about what features of songs whales attend to and what factors determine how and when singers change songs, it's premature to assume that any change will do as long as it is new.

Each of these symbols is one unit.

Different symbols indicate sounds that sound different.

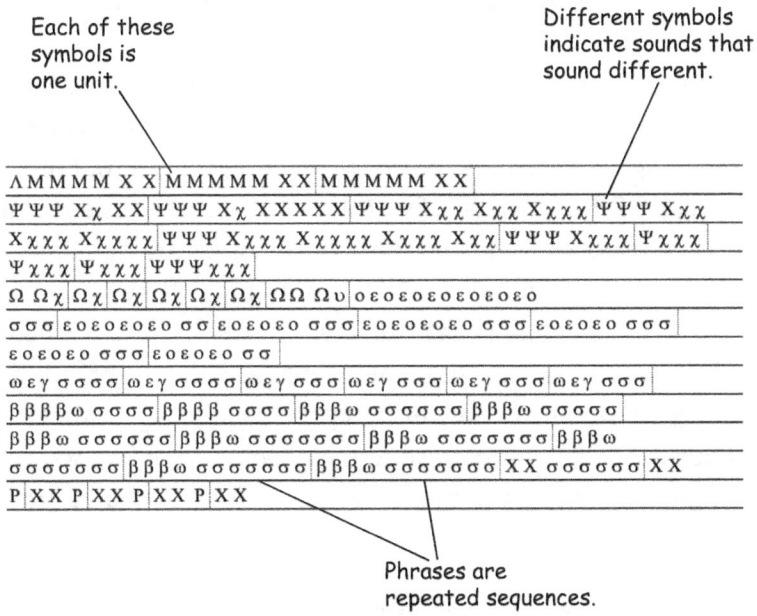

Phrases are repeated sequences.

Songs traditionally are transcribed as sequences of labels or symbols—one per unit. These sequences are then compared across years to see how much songs change from one year to the next. Song sequence transcribed by Winn and colleagues in their 1970 report.

change in different populations might vary depending on how densely singers congregate in specific regions.

Regardless of the mechanisms leading singers to continuously vary their songs within and across years, the fact is that adult humpback whales never stop modifying their songs. No other echolocating animal does this. Bats and toothed whales use a highly stable set of sounds to generate echoes. Critics of the sonar hypothesis have been quick to point this out, noting that "if song were used for sonar, one would expect that the process of natural selection would result in convergence to an optimal signal."[19] The problem with this critique is that the optimal signal for monitoring distant whales in a variety of oceanic contexts is not known. The sets of units that singing humpbacks use in any given year might indeed be optimal for this purpose. Many of the changes that singers make to songs involve stretching, pitch shifting, dropping, or duplicating units. Bats morph their sound sequences in similar ways depending on where they are echolocating and how close they are to targets. Singing humpbacks change their units and phrases more extensively than bats or dolphins, for reasons that are not yet clear, but those changes are not arbitrary[20] and don't appear to reduce the echo-generating potential of units across years. I don't know why singers morph the units in their songs. No one does. But it's possible the changes are beneficial for echoic perception, and at least they shouldn't disrupt a whale's ability to use songs as sonar. Also, while it's not yet possible to identify or explain the mechanisms that lead different populations of humpback whales to converge on the same song changes, the fact that they do so points to shared environmental factors, genetic constraints, or ecosystem changes that do not require singers

---

19. From Whit and colleagues, "Against the humpback whale sonar hypothesis," page 299.

20. And probably not innovative or erroneous, either.

to copy or improvise changes they hear other singers making. And the fact that songs change over time does not prevent units from producing echoes or singers from perceiving and acting on those echoes.

The dynamic nature of humpback whale song, with its constantly shifting units, phrases, and themes, was what led me into the field of neuroscience. There's no question that singers modify their vocal repertoire in ways unlike other cetaceans or terrestrial nonhuman mammals. That means that whoever is listening to songs and song echoes must constantly adjust what they are listening for over time. Humpback whales listening to songs are like audiophiles incessantly learning to recognize and appreciate new acoustic scenes. If singing humpback whales use songs as sonar, then they must be adaptively processing incoming echo streams in ways unmatched by any other echolocating animal. The only way singing humpbacks can do this is if their brains have highly sophisticated circuits for constructing echoic scenes. Not just any neural circuits will do. Singers must, in essence, be constantly learning to hear, and for that to happen, their auditory circuits must be changing all the time.

CHAPTER 9

Within Whales' Heads

THERE'S A BATTLE RAGING over the mental states of cetaceans. On one side are the skeptics: intellectuals who recognize that peoples' beliefs about the inner lives of whales and dolphins are based more on questionable assumptions than on hard evidence. The opposition consists of defenders of nature who are convinced that cetaceans have established their sentience and thus should be treated with greater respect and empathy.

Humpback whales are often poster children in this debate. It's not just because of their bewildering songs. Humpbacks can show an odd affinity for humans that borders on indulgence, especially for a wild animal once hunted to near extinction. Consider the case of Nan Hauser, a whale researcher at the Center for Cetacean Research and Conservation in the Cook Islands, who was "saved" by a humpback whale. Nan had entered the water near two male humpbacks as part of a filming project when one of the males approached her and started pushing her through the water. This was a precarious position for Nan since the whale might easily have injured her. At one point the whale lifted Nan out of the water with one of its pectoral fins.[1] Around this time, Nan noticed a large tiger

---

1. The fins that look like wings.

shark swimming beneath them—large as in 18-feet-long large. Tiger sharks are no joke. They're in the club of shark species most likely to chomp on a human, and they do not nibble. As Nan began to panic, the humpback lifted her out of the water with its head and carried her back to the boat.

From Nan's perspective, the curious whale was her savior. It's hard to know the whale's perspective of the situation, but other incidents suggest her interpretation is not far-fetched. For instance, groups of humpback whales have been seen protecting various animals from attacks by pods of orcas, including sea lions, seals, a mother gray whale and her calf, and even a sunfish!

Many scientists view such rescues as a side effect of humpback whales' natural tendency to protect the young of their own species. Others, however, think that humpbacks' defensive actions could indicate altruistic impulses combined with the physical power to act on those compassionate feelings. Also, the ability of humpbacks to counter the attacks of a group of motivated, large-brained dolphins[2] suggests some level of cognitive sophistication and possibly some understanding of the orcas' goals and strategies.

You don't need to take a side in this debate to acknowledge that things may be going on inside the heads of humpbacks that humans have yet to grasp. Understanding other species' minds is a challenging task in the best of circumstances. You can watch what animals do and take a shot at interpreting their actions, but that can only get you so far. Aside from behavioral observations, the main pieces of evidence that scientists use to determine what whales can do mentally are the structural and functional features of their brains. In an ideal world, it would be possible to examine a whale's brain and point to different features that indicate different mental capacities, as one might gauge an animal's biting power by examining

---

2. Orcas have larger brains than any other dolphins, reaching sizes that rival those of sperm whales.

Humpback whale

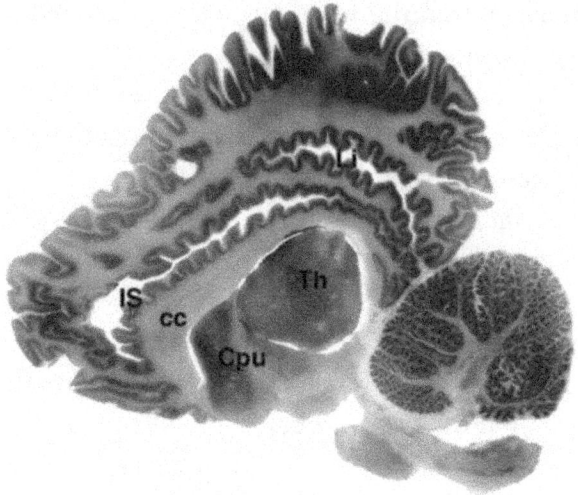

Dolphin

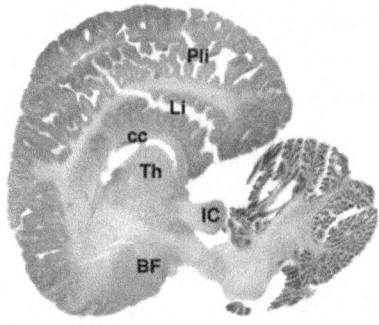

Cat

Hedgehog

A humpback whale's brain is just as complex as a bottlenose dolphin's brain and slightly bigger. Both brains contain all the basic structures seen in land mammals. The brains of most terrestrial animals (including cats and hedgehogs) are much smaller than those of whales and dolphins.

its teeth. Sadly, neuroscience has not quite reached that stage. A major roadblock to inferring mental function from brain structure is the wide variety of neurons that interact with each other in complex ways, making it difficult to identify what it is that networks of neurons do.

Debates about cetacean brain power are just as contentious as arguments about their altruistic actions. While many mysteries about whale and dolphin brains remain unsolved, a century of neuroscience research hasn't been in vain. More is now known about how mammalian brains work than ever before, making it possible to identify the kinds of neural mechanisms that make abilities like echoic perception possible. Neuroscientific studies of hearing in bats, birds, humans, rats, and monkeys can't reveal the sorts of thoughts and beliefs a singing whale might have, but they can clarify what needs to happen for a whale or dolphin to be able to see with sound.

(continued)

The cauliflower-looking parts of whale and dolphin brains are what is called the cortex, with the larger lump (*to the left*) making up the bulk of the brain, called the cerebral cortex, and the smaller lump (*to the right*) called the cerebellum or cerebellar cortex. Both structures are highly convoluted, which gives them a wrinkled appearance. These brain wrinkles happen when large sheets of cortical tissue are squashed into a restricted space, somewhat like stuffing a shirt into a tube. A humpback whale's cerebral cortex is a bit thicker than a dolphin's but otherwise is similarly organized. The convoluted cerebral cortex of humans is often pointed to as the home of all the cognitive processes that make humans special, which is why scientists like John Lilly were startled and excited to see just how much cerebral cortex was crammed inside a dolphin's skull. For some reason, the same sorts of convolutions in baleen whale brains provoked more of a scientific shoulder shrug. *Source*: Cetacean brain images reprinted with permission from Hof and Van der Gucht (2007).

Belittled Brains

I noted in chapter 2 how Orthello Langworthy gushed over the complex structural features of dolphin brains in the 1930s while at the same time describing their brains as hedgehog-like.[3] He also remarked on the enormity of dolphins' auditory nerves long before it was recognized that dolphins could echolocate. While dolphins' brains were being intensively studied and precisely characterized in the scientific literature, the first scientific description of a humpback whale brain simply noted that its various parts "conform to the typical cetacean pattern" (Breathnach, 1955, p. 343). It would be another 60 years before the second scientific description of a humpback whale's brain was published.

The scarcity of neuroanatomical studies of whale brains was not simply because they were perceived to be basically the same as dolphin brains. It was in large part pragmatic. Even today, getting your hands on a fresh baleen whale brain is a monumental task. Samples from washed-up carcasses are usually too decayed to reveal much since whales' brains deteriorate rapidly after death. Few other scenarios exist in which a scientist might encounter their brains. In the last 40 years, neuroanatomists have only analyzed about 20 baleen whale brains, most of which came from bowhead whales killed during hunts by Alaskan Eskimos or from minke whales killed by whalers in Iceland. Those analyses revealed more details about the neural circuits that make up baleen whales' brains, including information about the sizes and types of neurons located in specific brain regions, how those neurons are distributed, and how brain structure in baleen whales compares with what's seen in toothed whales. But

---

3. Cetacean brains are often compared to hedgehog brains because the general structure and collection of neurons within a whale's brain look more like a hedgehog's brain than a human brain.

inferring mental capacities from studies of brain anatomy is no easy feat.

The most obvious way to compare the brains of different species is by looking at their absolute size. More brain tissue could mean more thinking power. But as bodies get larger, their constituent parts—including brains—get larger too. And bigger brains don't always translate into more mental abilities. For instance, few chihuahua owners would concede that their pretty pooch (brain weight of 3-6 grams) is a hundred times dumber than a Saint Bernard (brain weight of 200-300 grams). If anything, they might judge the dog with the larger brain to be the lunkhead.

Many neuroscientists agree that there is no direct relationship between quantity of brain tissue and mental capacity.[4] To compare brains across different-sized species, scientists often use a measurement called the *encephalization quotient*, or EQ. EQ is calculated by dividing an animal's brain size by the expected brain size given the animal's body size.[5] EQs treat brains as if they were hands. Children have smaller hands than NBA players because they have smaller bodies. A child whose hands are extra large relative to his peers would still have tiny hands in comparison to any professional basketball player. In the same vein, higher EQ measures reveal when an animal's brain has relatively more tissue than the norm for similarly sized species. Some scientists argue that species endowed with more brain tissue than expected based on their body size are more likely to possess extra mental power, just like big-handed children would be more likely to be loud clappers.

As you might expect, humans have the highest EQ of any species. Dolphins and chimpanzees come in at a distant second and third.

---

4. For example, surgically removing half a child's brain through a procedure called hemispherectomy—a last-ditch approach to counteract debilitating seizures—typically doesn't reduce the child's mental abilities or academic achievements.

5. The expected brain size is based on the consideration of lots of different species of various sizes.

Given that humpback whale brains are only slightly larger than that of a bottlenose dolphin (as illustrated on page 214) while their bodies are much bigger, you can probably guess what a humpback whale's EQ is like—it's much lower.[6] Now you might be thinking, If humpback whales have a brain that's bigger than a dolphin's, shouldn't they be able to achieve similar mental feats? Therein lies the problem with treating brains like hands. While the fingers of an NBA player might be four times longer than the fingers of a five-year-old, structurally those digits are essentially equivalent. Brains, in contrast, are made of a whole host of different cell types and structures, many of which do not inflate like a balloon as an animal grows larger. In that respect, brains are more like computers. It's not the size of the processor that determines what a computer can do but the number, variety, and interconnectedness of components within that chip that matter. Similarly, it's probably not the absolute size of a brain, or even its relative size, that is important for mental abilities—it's more likely that the specific cells, structures, and connections within the brain determine what it can do.

From this "components matter" mindset, neuroanatomists have begun paying more attention to the number and variety of brain cells within different species of animals to identify which variations have an impact in terms of cognitive prowess. Neuroscientists tend to focus on those brain cells most closely associated with processing sensations and generating movements: neurons. Everything you perceive, think, feel, or do depends on neural activity. The same is true for whales and dolphins. Given the importance of neurons for brain function, it's natural that many scientists view the number of neurons within an animal's brain as indicative of its "brainpower."

---

6. The EQ of a human is about 7.0, while the EQ of a chimpanzee is around 2.5. The EQ of a bottlenose dolphin is approximately 4.0. The EQs of all baleen whales are below 1.0, meaning that their brains are smaller than expected based on their large body sizes.

New methods for counting neurons have revealed that the number of neurons a given species possesses is not easily predictable from measures of overall brain size. Different brain structures of similar sizes can contain vastly different numbers of neurons. For instance, an elephant's brain contains a whopping 257 billion neurons, but only 6 billion of those are in the cerebral cortex.[7] Human brains, in contrast, contain only around 86 billion neurons. But the human cerebral cortex contains about three times more neurons than that of elephants.

In addition to neuron numbers, scientists have also begun to study the different types of neurons that compose the brains of different species. Cetacean brains contain several types of neurons that aren't seen in human brains. Their brains are also missing some types that are common in humans. Neurons within a humpback whale's cerebral cortex are bigger and branch out more extensively than is typically seen in dolphins and most land mammals. How these structural differences affect brain function are unknown. Despite clear similarities to the brains of land mammals, whales' brains are sufficiently different from other mammalian brains that neuroanatomists suspect cetaceans may have evolved unique mechanisms for transmitting signals across networks of neurons.

Scientists are still discovering the variety of neurons and other cells within brains. In 2023 neuroscientists identified over 3,000 different cell types in the human brain alone, with some types only found in specific brain regions. It's not yet known how many neurons there are in any whale's brain or whether their brain cells are as varied as those of humans. Even if such details were known, it wouldn't necessarily reveal whales' mental capacities or how singing whales use songs because of the unique features of whales' brains. Until neuroscientists know more about how neural circuits

---

7. The remainder are in the elephant's cerebellum—the "little brain" hanging beneath the cerebral cortex, near the back of the brain.

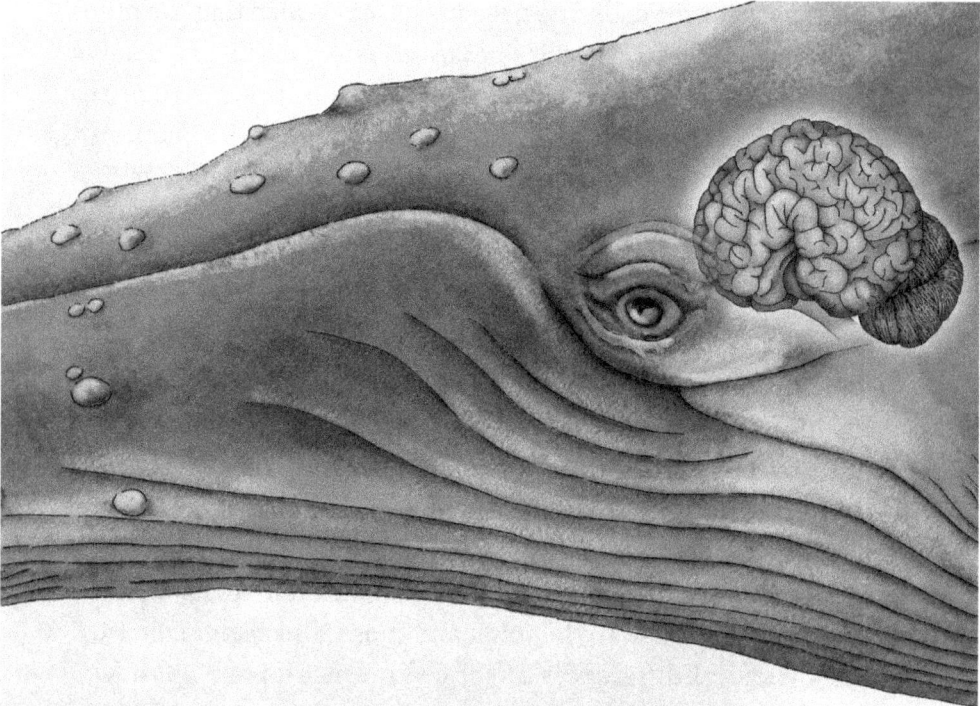

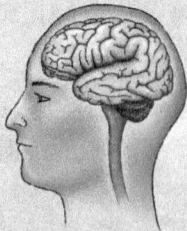

work, it's hard to predict which of the structural features of brains determine mental capacity.

What could help, though, would be more details about how whales' neurons are interconnected. That's because neuroscientists know a lot about specializations in the brains of bats that enable them to interpret incoming echoes. The sonar hypothesis predicts that whales' neural circuits have evolved in ways that enable them to perform similar perceptual functions to echolocating bats. In the same way that evolutionary forces shaped bats' wings to look like birds' wings, the brain circuits possessed by an echolocating whale could resemble those of bats.[8]

Over the last 60 years, neuroscientists have probed the brains of echolocating bats. One of the main tools they've used is electrophysiology—basically, recording the electrical activity of specific neurons in response to incoming sounds. Electrophysiological experiments have shown how bats' brains are specialized to process and interpret echoes. Neurons connected to a bat's cochleae constantly respond to sound-induced vibrations. Sounds that are sufficiently loud or interesting typically trigger a volley of activity that travels from neuron to neuron within the bat's brain. Although echoes from its own cries are seldom loud, they are some of the most interesting sounds a bat hears. As waves of neural activity triggered by echoes propagate through a listening bat's brain circuits, there's a kind of spatial sorting process such that certain types of echoes lead to increased activity in specific parts of the brain. In this way, patterns of neural responses from the cochleae

8. Though long-range, underwater sonic echolocation is sufficiently different from short-range ultrasonic echolocation in air that one would expect whales' neural circuits to show some unique specializations.

Opposite page: *Brainy mammals.*
Humpback whales possess large brains with an impressively convoluted cerebral cortex. What they can achieve perceptually and cognitively with their massive brains remains unknown.

are transformed into categories of objects such as edible versus inedible, small and moving versus big and stationary, or catchable versus a waste of time.

In addition to neurally sorting sounds by types, bats' brains can also sort them along specific dimensions. Imagine you're given the task of sorting buttons. You could sort them into piles of those you like versus those you don't. But you could also sort them based on their color, size, weight, or value. These are dimensions along which buttons can vary continuously. Neural activity patterns in a bat's brain also vary continuously in response to echoes, depending on changes in a target's distance, speed, height, and orientation.

The most sophisticated sorting of neural activity occurs within a bat's cerebral cortex. There, neurons respond most vigorously to specific combinations of produced sounds and resulting echoes, allowing neuroscientists to observe the aspects of incoming sounds that cause the strongest reactions in different parts of the cerebral cortex. For instance, certain cortical neurons respond most to pairs of sounds and echoes separated by specific time intervals (called *time delays*), which vary based on a bat's distance from a target. Time delays corresponding to distances between 30 and 600 centimeters[9] tend to trigger a lot of activity in a bat's cerebral cortex. Hunting bats are known to home in on dinner at distances within this range.

In some species of bats, neurons sensitive to target distance are spatially organized within the cerebral cortex. For instance, neurons that respond most to distant targets might be clustered within a leftward cortical region, and neurons that react vigorously to nearby targets might be found to the right of the far-loving cluster. With this kind of organization, a neural activity wave will sweep from one part of the cerebral cortex to another as a bat closes in on its prey.

---

9. About 1 to 20 feet.

Of course, not all echoes a bat hears will be self-generated. When multiple bats are echolocating in the same area, they must give greater weight to their own echoes if they want to correctly interpret what's happening in front of them. Cortical neurons are particularly sensitive to echoes received immediately after a bat has produced an echolocation cry. When an echo is detected within this expected time period—called an *integration time window*—neurons shut down so that other echoes don't trigger additional activity. In this way, bats can filter out some of the calls and resulting echoes produced by other nearby bats.

No comparable electrophysiological recordings have been made from the brains of baleen whales. Nevertheless, cortical organization in echolocating bats provides important clues about the kinds of neural sensitivities that should be present if singing whales use units as sonar signals. Specifically, for singing whales to track other whales using echoes, they would likely need cortical neurons that are sensitive to spatial dimensions corresponding to distance and direction. Given that bats process echoes generated by ultrasonic, ultrarapid "units," the same kinds of neural sorting mechanisms seen in bats could potentially work for baleen whales.

Based on findings from the brains of echolocating bats,[10] one would expect whales' cerebral cortices to contain large numbers of neurons that are triggered by specific combinations of echoes occurring immediately after produced units, with an emphasis on faint echoes coming from distant targets. Some of those neurons should be sensitive to the timing of echo arrivals, while others should be tuned to fluctuations in specific pitches. Assuming whales' brains sort units and their echoes in ways that parallel

---

10. The bat species that uses the widest variety of echolocation cries (think little goblins) probably makes the best model for what to expect from humpback brains. Electrophysiological studies of the most versatile bats haven't happened yet, unfortunately.

what's seen in echolocating bats, they should have separate corti-cal regions that react to variations in how far echoes have traveled and to differences in the directions from which echoes arrive.

The main complication in extrapolating from the neural pro-cessing of echoes by bats to that by singing whales is time. The rapidly streamed sonar signals of bats match well with the rapid rates of neural firing in their cerebral cortices. The integration time window necessary for a singing whale to monitor relevant echoes would have to be much longer—maybe 100 times longer than that of an echolocating bat. Monitoring such long intervals would require neural circuitry beyond what's present in bats. Human brains can integrate auditory information over relatively long intervals when processing music and speech, however. So lon-ger integration time windows are within the range of what mam-malian brains can achieve.

Many of the neural circuits necessary for echolocation are spe-cialized for processing detailed features of incoming sounds, but these are not the only adaptations seen in echolocating animals. Another capacity that many echolocating species share is vocal learning—when an animal modifies its vocalizations based on ex-perience. Flexible vocal learning is rare among mammals, found only in a handful of non-echolocators.[11] Yet many species of ceta-ceans and bats have the ability to dynamically adjust their vocal-izations based on sounds they hear. They can accomplish this in a couple of ways. First, they can learn which sounds to produce by hearing examples of those sounds, like you do when you start saying "cheugy" after hearing some drip Zennials use the term. Echolocators may also learn to adjust their echolocation cries de-pending on the acoustic context.

Bats are highly flexible when it comes to selecting and morph-ing echolocation cries, but it's not clear how much learning

---

11. Including humans, elephants, and some seals.

contributes to their vocal variability. So far, there's no evidence that bats imitate any bat slang they overhear.[12] Unlike bats, toothed whales show less flexibility in their production of echolocation signals. But they do sometimes learn to imitate weirdo sounds that they hear humans making. The fact that only a few mammals demonstrate flexible vocal learning suggests that some "extra" neural hardware may be required—specifically, circuits linking auditory processing to vocal motor control systems.

It might not be immediately obvious why the ability to imitate or modify vocalizations would be either rare or linked to echolocation. Most mammals just make the sounds they were born to make: The mouse goes "squeak," the cow goes "moo," the cat goes "meow," and the fox goes "ha." If a few species evolved more flexible vocal control options, then presumably it's because their ancestors benefited in some way from this innovation. While many scientists are studying how the extra neural circuits needed for vocal imitation evolved in certain mammals, the question of why imitators are often echolocators has gotten less attention. Intuitively, it makes sense that producing dynamic streams of echolocation signals requires some fancy vocal control mechanisms that tap directly into what the vocalizer is hearing. But what's the evolutionary advantage of making this kind of sophisticated system depend on learning?

A possible answer to this question comes, surprisingly, from fee-beeing chickadees. As I noted in chapter 7, the fee-bee song of the chickadee is made up of just two notes: the ding-dong of birdsong. Despite its seeming simplicity, chickadees learn to produce this song by hearing other chickadees sing it. Scientists know this from

---

12. Bats do seem to modify the pitches they produce based on what they hear other bats doing, however. Although bats of the same species produce highly similar echolocation cries, different groups can demonstrate significant variations in the pitches they use. When bats move from one roosting site to another, they often adjust the focal pitches of their echolocation cries to match those being used by the bats in their new home.

experiments in which chickadees are raised in isolation. Chickadees that grow up without hearing proper songs sing them incorrectly.[13] Why would evolution favor song learning in chickadees when so many other bird species produce fancier songs without the need for learning? It's probably not so that singing males can show off their vocal smarts to prospective mates since all male chickadees sing the same mundane song.

You may recall from chapter 7 that the chickadee's fee-bees produce an interesting kind of reverberation that can reveal to listeners how far the song has traveled. I also noted that humans are better at judging the distance to a sound source when the sound is familiar. In the 1980s, ecologist Eugene Morton proposed a hypothesis to explain why some birds might learn songs. His idea, which he called the *ranging hypothesis*, was that if a bird can reproduce a song it's heard, then it can compare its memory of its own song to degraded versions of songs heard from different distances, enabling the imitator to judge the positions of other singers more accurately.

It works like this: Imagine you've been assigned the task of picking up a stranger at an airport. If you have a close-up photo of the person, it will be much easier for you to spot that individual in a crowd from a distance, even though the person's face may look more disheveled or be partly obscured by other heads. In a similar way, a bird can potentially compare what their own songs sound like to what other birds' matching songs sound like when produced from far away[14] to estimate their distance from those other singers. In this scenario, a bird hearing a song that it cannot produce itself will be less able to assess where that unfamiliar song originated.

---

13. Specifically, isolates produced more than two notes, and the differences between these notes were not normal.

14. Generally, songs perceived after farther propagation will be much quieter, with some parts potentially obscured or otherwise distorted.

Neuroscientists have found that certain neurons in songbirds' brains are especially responsive when a bird hears its own songs, consistent with the idea that the songs a bird produces influence how it hears songs. Whether these selective neural reactions affect a bird's ability to judge the locations of other singing birds has yet to be tested. Intriguingly, humans are better able to judge the distance of a loudspeaker playing back normal speech compared to one producing speech played backward, suggesting that humans, at least, are better at localizing learned vocalizations that they themselves can make.

What does the ranging hypothesis have to do with the relationship between echolocation and vocal imitation? An animal with the ability to reproduce new sounds it hears can, in essence, create a vocal selfie that it can then use to better localize sounds coming from long distances. And this is true regardless of whether those matching sounds are vocally produced by another animal or are bounced back from an animal in the form of echoes. Vocal imitation may be common in echolocating species because both imitation and echolocation require comparing produced sounds to perceived sounds, and both may serve to enhance an animal's ability to spatially hear.[15]

Unlike songbirds, cetaceans can keep track of others' movements either by actively monitoring them (through echolocation) or by passively localizing the various sounds they produce (like a human might). You can view ranging as a hybrid of these two basic listening processes. As in echolocation, the listener compares incoming sounds to memories of its own vocalizations to apprehend where the sounds originated. From this perspective, the

---

15. Vocal imitation abilities are evident in species that do not echolocate, suggesting either that echolocation is an adaptation that builds on existing vocal flexibility or that the conditions that favored the evolution of echoic perception differed somewhat from those that favored the evolution of auditory-motor specializations for passively localizing sound sources.

main difference between ranging and echolocation is the type of memory that the listener is using as a basis for comparison: An echolocator uses memories from recently performed actions, while a ranging listener relies more on long-term memories of all the sounds it's capable of producing.

It's clear that both whales and dolphins benefit from being able to hear where their comrades are located. Keeping track of who is where using vision alone is rarely an option in ocean habitats. The only way that cetaceans can hope to coordinate their movements is through listening and localizing. Vocal imitation abilities can make this easier. It may seem odd to think of a copycat as someone who is squinting their ears to better track the locations of others. That's because you, as a primate, mainly rely on vision for that sort of thing. For orcas hunting in the night or lone humpbacks singing into the abyss, hearing the actions and movements of others is often the only viable option.

If vocal imitation in cetaceans serves as a hearing aid, this has important implications for interpreting the actions of vocalizing whales and dolphins. For instance, in chapter 8 I described the signature whistles that dolphins often produce when they're isolated in tanks or far from other companion dolphins. Dolphins are known to imitate the signature whistles of their mothers from a relatively early age. Young dolphins spend most of their calfhood years close to their mothers, but at times a calf will head off on its own to explore the world. Calves may swim 50 meters or more away from their mothers, ranges at which visible cues would not reliably reveal the calves' (or mothers') locations underwater. Researchers discovered that when mothers and calves are separated in the wild, impending reunions are often signaled by a profusion of signature whistles.

Perhaps you're envisioning a scenario in which Momma Dolphin gets worried about where little Jerky has gotten off to and starts calling out Jerky's signature whistle to let her calf know that it's time to head home. An endearing reunion tale, but totally wrong. Researchers found that before a reunion, the calf often repeated its own signature whistle multiple times and sometimes continued to do so until it reached its mother. The mother was typically a passive partner in the reunions, producing no whistles[16] and changing her behavior little as her calf approached. The whole sequence is odd. If the calf is calling to its mother, then why is the mother not responding? And if the calf knows where the mother is located and that she is unlikely to approach or respond, why is the calf bothering to whistle at all?

In one experiment involving captive dolphins, trainers instructed a mother dolphin to retrieve her calf from a distant location. The mother often repeatedly produced her own signature whistle after being given the instruction, but the calf showed no obvious responses to her whistles. Instead, the mother swam over to the calf's location—typically while whistling—and then brought her calf back with her to the trainer. Again, why did the mom produce her own signature whistle rather than the calf's? And why did the calf ignore her mother's signature whistles?

A mother dolphin can almost certainly perceive that her calf is homing in on her position when the calf repeatedly whistles while approaching. According to the ranging hypothesis, the mom's ability to imitate her calf's signature whistle should enhance her ability to monitor the calf's movements through listening. Repetitive whistling by a calf during a reunion will inform other dolphins about the calf's location, where the calf is heading, and, potentially, its intentions.[17] If the mother were to whistle back to the calf, this

---

16. In some cases, mothers have been observed to stop moving when the calf is approaching, which could be evidence that she is waiting for the calf to reach her.

17. Note that this information is also available to individuals that are known to kill calves, such as orcas and some male bottlenose dolphins. Signature whistles can travel 1,000 meters or more, providing ample opportunity for eavesdroppers to detect the presence and location of a vulnerable calf.

would provide the calf with the same kind of spatial information. But does the mother need to whistle back? Or might the calf already have the relevant spatial information? Recall that there are two ways in which a dolphin (or whale) can potentially reveal its location: either by making sounds or by reflecting sounds produced by others. If signature whistles reflect from the lungs of dolphins,[18] then a calf's ability to vocally imitate should enhance its ability to localize its mother using those reflections. In other words, dolphins' vocal imitation abilities might enable them to use their whistles as sonar signals in much the same way that I'm arguing singing whales use their songs.[19]

Many researchers claim that song learning by baleen whales evolved for radically different reasons than the evolution of whistle imitation by toothed whales. For instance, biologist Vincent Janik argued that vocal copying evolved in toothed whales to facilitate individual or group recognition but suggested that song imitation in baleen whales evolved through sexual selection. The ranging hypothesis proposes that chickadees, baleen whales, and dolphins may have all evolved the capacity to imitate sounds because this improved their ability to localize other vocalizers. And since most cetaceans benefit from monitoring the movements of other members of their species, it's likely that most cetaceans learn at least some vocalizations by hearing and copying them. This may even apply to those cetaceans that produce the simplest of songs.

In the world of baleen whales, blue whales are the kings of simple songs. Their songs typically consist of one to three units. Units in blue whale songs can last from 10 to 20 seconds and are produced every 1-2 minutes. They are much lower in pitch than the units

---

18. A possibility that has yet to be tested.

19. In which case scientists might want to consider whether it's more appropriate to refer to dolphins' signature whistles as "singnature whistles."

produced by singing humpback whales. When you think blue whale song, think low and slow. Blue whales from specific geographic regions sing characteristic songs that are so predictably structured that they can be used to identify a singer's home range. Unlike humpback whales, singing blue whales do not constantly morph their units or vary how they produce units in patterns across years.

At first glance, singing blue whales seem like slow-motion ocean donkeys, producing a stereotyped, whale-scaled "hee-haw" over and over throughout their lives. There's little reason to suspect they would need to learn their songs. The differences in songs across populations could easily result from genetic variations. Many other mammals, like howler monkeys, produce more impressive songs without learning them. After many years of recording blue whales singing their simple songs, however, researchers noticed an odd trend in the recordings. Whales seemed to be gradually lowering the pitches of their units every year. The changes in pitch were tiny, but over many years the shift accumulated so that when researchers plotted them across decades, a clear and steady descent could be seen. Blue whale songs were sinking.

Even more surprising, this downward shift was evident in all populations of blue whales, regardless of the differences in their songs. I know you're wondering why this might happen—so are the scientists studying blue whale songs. Many explanations have been proposed and subsequently shot down. One possibility still in the mix is that blue whales are adjusting their songs based on what they hear other singers doing. Proponents of this hypothesis point (predictably) to sexual selection as the force driving singing blue whales to shift, echoing experts' explanations for why humpback whales progressively change their songs. The blues-copying-blues hypothesis provides a possible explanation for why songs might change over time. But saying that singing blue whales go down because female whales like it is not so scientifically satisfying. It's a hypothesis that's virtually impossible to test, which is the least useful kind of scientific hypothesis.

Sex is not the only thing that might cause a whale to sing low. Echolocating bats also shift the pitches of their cries to match those of their new neighbors when they switch to a new roosting site. Might blue whales be doing something similar? While thinking through this possibility, I decided to analyze a few blue whale songs myself. I was mainly curious to see how exactly blue whales shifted units in their songs. I had no intention of formally analyzing the songs.[20] When I started looking closely at the pitches in blue whale songs, I got this eerie feeling of familiarity. It wasn't because the blue whales were singing phrases like those used by singing humpbacks, though. Instead, the blue whale songs seemed weirdly like chickadee fee-bees. At first, I thought this was probably because I'd spent too many years staring at spectrograms of fee-bees. But when I measured the pitches, I found they were related to each other in the same way that fees are to bees within a chickadee's song. Specifically, there is a pure tone followed by a second tone with a pitch that is nine-tenths the value of the first tone. It was the closest convergence I had ever seen between sounds produced by whales and birds. Blue whale songs are what you get if you take a chickadee, inflate its size by a factor of 20 million, and then make it sing underwater. Blue whales even transpose their songs like chickadees, proportionately decreasing all the pitches within a song so that the relationship between the tones stays the same as the song shifts.

The fee-beeish nature of blue whale songs means that nonvocal blue whales—like chickadees—can potentially use song-generated reverberation to judge their distance from far-off singers. It also suggests that however chickadees benefit from learning their

---

20. You'd be surprised how possessive bioacousticians can get about the sounds of their favored species—only a few elites have dared to analyze the songs of multiple species of baleen whales.

simple songs, blue whales would likely achieve the same gains from learning theirs. The ranging hypothesis suggests that what both blue whales and chickadees stand to gain from learning songs is enhanced spatial hearing. Importantly, in the case of blue whales and other baleen whales that change their songs over time, this process of auditory enhancement is not limited to early development. Since whales constantly change the units within their songs from one year to the next, the only way they can retain the kinds of advantages that the ranging hypothesis proposes is if their capacity to detect and recognize specific song features also changes in parallel. Unlike most other mammals, singing whales may need to keep up with the latest sonic trends to maintain their perceptual prowess. Regardless of how whales use songs, the only way they can maximize their ability to detect, recognize, and differentiate units that constantly change is by continually adjusting how they process incoming sounds.[21] And the only known way that whales, or any other organism, can constantly adjust how they process sensations is by modifying the underlying neural circuits.

The Ever-Changing Cells of Whales

Classical models described sensation and perception as if they were the natural outcome of cascading neural reactions to events in the world: Physical energy hits biological receptors, triggering neural activity that passes through tiny tissue channels to be passed on to more neurons, further spreading the word throughout the brain, and voila!—you hear sounds. In this biological bucket brigade model, what you hear is what you get. Just as pianos respond in predictable ways when specific keys are hit in specific combinations,

---

21. The sonar hypothesis predicts that singers' constant morphing of units within and across songs is in some way perceptually advantageous to singers, perhaps by decreasing neural adaptation or reducing cross-song interference. It's unclear why these benefits would only be helpful to humpbacks and not other whales, however.

MOVING BEYOND THE SONGBIRD SNUB

Scientific comparisons between singing whales and singing birds was a good idea—in the 1960s. But like other "good ideas" from the sixties (think asbestos and lead paint), the portrayal of whales as oceanic songbirds has outlived its usefulness. It's time to broaden the scope of cross-species comparisons to include animals that face similar perceptual challenges to those faced by baleen whales.

I'm not opposed to all bird-whale comparisons, just like I'm not opposed to all kinds of paint. Singing chickadees taught me new ways that vocalizing animals can use reverberation that I would never have discovered on my own. But if scientists are going to continue fixating on birds as the best possible models of baleen whale vocal behavior, then they should focus some of that effort on birds that search for small movements over long distances, such as eagles and hawks.

Bald eagles can detect the movements of submerged fish from hundreds of feet away, and golden eagles can detect a running rabbit from one to three miles away. Rabbits are fast movers, but when viewed from such long distances they probably look like gorged fruit flies slowly turtling themselves away from a watermelon feast.

Eye anatomy and visual scanning behavior have been studied extensively in raptors to understand how they are able to detect and track tiny targets moving within such complicated vegetative cover. Identifying the adaptive specializations in raptors' visual perception that make high-resolution, long-distance movement tracking possible could go a long way toward clarifying the kinds of specializations that large whales need to successfully perform similar search tasks using sound.

And why limit cross-species perceptual comparisons to birds? Mountain sheep are thought to have amazing visual sensitivity to the movements of distant encroachers, which is why they are challenging game for hunters and other predators. In the spirit of John Lennon, all I am saying is give sheeps a chance. Orcas and bats are probably comparably sensitive to the movements of distant prey, but no one really knows how far away they can echoically pick out future meals.

Treating humpback whales like sea canaries minimizes their magnificence. Just as raptors are lords of the sky, surveying large swaths of nature every day of their lives, singing whales are sea sovereigns, vigilantly overseeing vast stretches of the waters surrounding them. They deserve to be recognized for what they are—gods of perception, not blubbery yodelers.

brains were thought to generate the same experiences when sections of cochleae were vibrated in specific patterns. If individuals' hearing experiences differed, this was attributed to genetically specified differences in the run of dominoes leading from that individual's cochleae to their cerebral cortex. Or maybe there were some weak "tiles" in the run caused by damage or disease. Within this classical framework, once a whale reaches maturity, its hearing hardware is set in place. As long as the whale stays healthy, it will hear sounds in accordance with whatever sensitivities it inherited from its parents.

This is the model of brains I had in mind when I first started studying the songs of humpback whales. According to this classical view, humpbacks have platoons of neurons that react in predictable ways whenever a specific unit[22] impinges on their

---

22. Or phrase or song.

heads. Engineers prefer these kinds of models because if you know what's going in, you know what's coming out. Things get trickier, however, when the incoming sounds are not pristine. For instance, imagine you are a female whale trying to compare two singers, one that is 500 meters away and another that is 2,000 meters away. Even if the singers produce identical units—which they do not, ergo the need to compare them—the properties of the units hitting your whale head will not be identical. That's because propagation will distort the two singers' units differently. You're going to need to account for that if you want the comparison to be fair. You can't risk choosing a loser just because they're closer! Accounting for such differences will require some additional neural hardware of the sort proposed by the ranging hypothesis.

Adding additional neural hardware would solve the problem of comparing singers in cases where they're singing the same units. But what if they're singing different units? Or even worse, singing some unit you've never heard before? What if they're constantly morphing their units and also changing how they morph them from one month to the next, like singing humpback whales do? Then, you've got a problem. Attempting to understand how listening humpback whales could solve this signal-processing nightmare led me to become a professor in behavioral neuroscience.

In typical engineering fashion, my initial approach centered on trying to find a comparable problem that someone else had already solved. While trying to solve the mystery of how listening humpbacks compare units, I discovered that modern neuroscience had progressed beyond the bucket brigade model. The new view was that mature sensory systems do not function like evolutionary Rube Goldberg machines. Instead, they constantly adjust in response to external events. Experiments with monkeys and rats revealed that simply having animals practice making distinctions between sounds or teaching them to associate sounds with specific consequences was enough to change how their cortical neurons reacted to those sounds. Neuroscientists also found that there were more connections from neurons in the cerebral cortex down to earlier stages of sensory processing than there were connections coming up from sensory receptors. Together, the new evidence indicated that cortical neurons do not simply react to sensory inputs. They shape what inputs make it through. And the way they do this is dependent on an individual's experiences.

The capacity of brains to show long-lasting changes in neural reactivity over time is referred to as brain plasticity. When those changes occur in circuits that react to cochlear inputs, it is called auditory plasticity. And when the changing circuits are in the cerebral cortex, it's called auditory cortical plasticity. One of the weird aspects of auditory cortical plasticity is that the neural changes it makes possible happen inside you every day. You are totally oblivious to most of those changes. That's because you're only consciously aware of your ongoing auditory experiences. If those experiences are subtly different today from what they were yesterday, you're unlikely to notice.

All mammals that have been tested, including echolocating bats, show similar auditory cortical plasticity. Although scientists haven't yet observed brain plasticity in any whale or dolphin, the neuroanatomical similarities between cetaceans and other mammals imply that cetaceans' cerebral cortices are at least as plastic as those of other mammals.

What happens, then, in the brain of a humpback whale that experiences a barrage of constantly varying song units? Almost certainly, their brains change as their exposures to units accumulate. At the very least, their brains change in ways that enable them to distinguish units or songs that are familiar from those that are not. Scientists know this from playback studies in which broadcasts of familiar songs consistently provoked nearby singers to approach the underwater speaker, while playbacks of foreign songs[23] repulsed them. Unless humpback whales' brains are unique in some yet to be discovered way, repeated exposures to songs will change how their neurons react to units within those songs. And since humpback whale songs never stabilize in structure or content, neither will the auditory cortical circuits of adult whales, regardless of their age.

Few mammals naturally modify their vocal repertoires as adults. Those that do usually add to an existing set of sounds, rather than replace large numbers of the sounds they use. In this respect, humpback whales differ significantly from most mammals, including most other toothed and baleen whales. Consequently, humpbacks listening to other singing humpbacks, or to echoes generated by their own songs, face special challenges relative to other echolocators. Why would humpback whales evolve to use songs that place such high demands on their neural circuits? An acoustic peacock tail does not need to constantly change. Most mammalian reproductive acoustic displays do not change at all, aside from changes associated with a vocalizer's growth. Appealing to sexual selection as an explanation for "why they do it" always carries an implicit addendum, never stated aloud by cetologists: Female preferences drove the evolution of singing by humpback whales *in ways that have never happened in any other living mammal*. This hidden addendum makes invoking sexual selection as a cause of humpbacks' singing dynamics little more than a restatement of the fact that singing humpbacks are doing weird stuff for reasons unknown.

The sonar hypothesis explains how humpback whales get away with constantly changing the properties of their songs by proposing that singers are both the sender and the intended receiver. Unlike passive listeners, a singer does not need to wonder what its own song was like before the song traveled multiple miles. It heard it.

---

23. Recorded from a different, noninteracting population of singers.

The sonar hypothesis does not explain, however, why a singing humpback whale wouldn't be better off using a fixed set of units. Echolocating bats morph their cries on the fly in ways that are reminiscent of singing humpbacks, but no bats progressively (or revolutionarily) modify the sets of cries they morph from one year to the next. This difference between whale songs and the sonar signals of bats and dolphins is an anomaly that critics of the sonar hypothesis were quick to jump on as evidence against its plausibility: "Because song is plastic and changes between seasons, shouldn't we consider every singer to be continuously 'learning to echolocate'? Then, taking this argument to its logical end, how could a male that uses a continuously inefficient sonar system out compete (in both proximal and evolutionary terms) a male that uses a reliable stereotypical sonar system?"[24]

Whit Au and his fellow defenders of the faith considered the possibility that any animal might continuously learn to echolocate patently absurd. Modern neuroscientific evidence suggests, however, that every echolocating species is continuously learning to echolocate in the sense that members' auditory cortical circuits are constantly adjusting to changing conditions and past experiences. Neither bats nor dolphins are robotic signal processors, as convenient as that would be for the scientists studying them. The key to understanding why humpback whales' songs are so dynamic may lie in Whit and friends' second critical query. Specifically, the logical argument they attempt to make presumes that any sonar system that constantly changes is, as a result, "continuously inefficient." It's true that if any existing human-made sonar started producing different sounds every time you used it, that would probably mean it's broken. And it's also true that toothed whales and bats use highly predictable sets of sounds when they

---

24. From Whit's epic rebuttal paper, "Against the humpback whale sonar hypothesis," page 299.

echolocate. But neither of these facts implies that a set of sonar signals that is not stereotyped is inherently inefficient or bad.

Here again, bat behavior sheds some light on what's possible. The little goblin bat has evolved ways of searching with echoes that differ from those of the perching horseshoe bat (which tunes into reverberation from a single tone) and the more ubiquitous big brown bat (which flexibly adjusts its signals to hunt insects on the wing). Each species of bat has evolved different echolocation strategies that work within specific foraging contexts. Now imagine a bat species with a mutation that compels it to travel all over the world—a cosmopolitan species that's always on the move. What sort of echolocation system would be efficient for this bat? One option might be: all the above. A bat that could shift its vocal repertoire to match its circumstances might arguably be a more efficient echolocator than one stuck using a single set of signals for all occasions. There's a reason that all bats did not evolve the same set of sonar signals. It's because the efficiency of any set is situation dependent.

If you apply this bat-based maxim to the case of singing humpback whales, you might conclude that it is sometimes a more effective strategy for an echolocating species to constantly modify its sonar signals than to use a fixed set of signals. In that context the key question is not why singing humpback whales change the sets of units they use in songs. It's *how* singers change the units they use and how those changes impact the echoes they receive.

Unfortunately, there's no simple answer. After two decades of working through the various possibilities, I've yet to reach the kind of epiphany that would bring all the puzzle pieces together in a satisfying way. I know a lot more about how singers vary the timing, pitch, and form of units but have yet to discover the rules of the change game or the pros and cons of different combinations of unit features. Within all sets of units used by singing humpback whales, some things change, and some things stay the same. There is a kind of holistic uniformity to any set of units used by singers in any given year that makes the units immediately identifiable as a set of humpback whale song components. You can see family resemblances across units from different years, such that most can be reliably sorted into "chirps," "screams," "moans," "grunts," and the like. You could say that all units within humpback whale songs are orbiting around some set of universal prototypes that never quite come into focus. These hidden prototypes are like Plato's Forms—each unit captures aspects of a Form but ultimately is a mere shadow of some underlying and frustratingly unknown essential nature.

My team, along with other scientists, discovered that the changes humpbacks make to songs each year are not as free-form and irreversible as they once seemed to be. There appear to be underlying "rules" about how units and phrases can change, both within song sessions and across days. Those rules could, in principle, relate to core properties of songs that need to be present, including physical limitations on the kinds of sequences that are most likely to bounce off large targets from far, far away. Detailed analyses of songs have also revealed that singers voluntarily control many different features of units. The changes that singing humpbacks make to units are *not* arbitrary in the way that trending fashions seem to be. They follow more predictable paths.

What's known definitively is that singing humpback whales change properties of their songs—both units and phrases—throughout adulthood. What can be safely inferred is that humpback whales' auditory cortical circuits change in parallel with song changes. What's not known is how these changes, both vocal and neural, relate to the functions of songs. The changes may increase the reproductive prospects of some singers and listeners, as proposed by the sexual advertisement hypothesis. Or, per the sonar hypothesis, they may enable singers to monitor their surroundings more effectively. For instance, these changes may reduce the likelihood of cross-echo interference.[25] The changes might even serve both purposes.

Nothing about the annual changes to humpback whale songs precludes units from being used as sonar signals. The fact that such changes occur does suggest, however, that if humpback whales use their songs as long-range sonar, then they've evolved echolocation strategies unlike those of any other echolocating animal. Some scientists view this as a weakness of the sonar hypothesis. If it is a weakness, it's one that the sexual advertisement hypothesis shares, for reasons noted above. I think it's more feasible to assess whether the broad range of units produced by singing whales creates useful echoes from distant whales than to assess whether some female humpbacks prefer certain units or combinations of units over others. In that respect, among others, I find the sonar hypothesis preferable.

Another outstanding question is: What makes humpback whale brains so much more receptive to everchanging inputs than the brains of other mammals? Is there something special about humpback whale brains that makes them more like a child's brain, ready to soak up new experiences in bulk? Hints to the answers come in part from recent investigations of sleep systems in cetaceans.

I noted in chapter 2 that cetaceans are among the few mammals that remain awake their entire lives. They achieve this by "sleeping" with half their brain while the other half remains alert, a phenomenon called *unihemispheric sleep*—a neat trick if you can pull it off. Neuroscientists interested in how whales and dolphins manage this feat closely examined circuits in these animals' brains known to regulate sleep/wake states in other animals.

---

25. Echoes that share acoustic features are generally more difficult to neurally separate than echoes with distinctive features and thus are more likely to interfere with a singer's ability to localize a target.

Coincidentally, one of the regions involved in sleep in mammals, the basal forebrain, also plays an important role in auditory cortical plasticity. Neuroscientists established the role of the basal forebrain in auditory cortical plasticity through a combination of behavioral and electrophysiological studies of rats. They found that if neurons in an adult rat's basal forebrain are artificially forced to activate every time a specific sound occurs, neurons in the rat's auditory cortex rapidly change their sensitivities to respond more vigorously to that sound. Auditory cortical neurons in bats were also shown to change in this way.

Studies of the basal forebrain in both whales and dolphins revealed that its structure was consistent with what's been seen in other mammals. This suggests that auditory cortical plasticity in cetacean brains is probably comparable to what's been observed in rats and bats, though likely not identical. When rats and bats learn about sounds, both of their hemispheres are in an awake state. In principle a humpback whale might be able to sing and hear songs using a single hemisphere. It's unclear how this might impact its auditory circuits. Perhaps each hemisphere retains its own personal history of sounds that arrived when the other hemisphere was snoozing. When you listen to speech versus music, the hemispheres in your brain will respond asymmetrically such that one side becomes more engaged than the other depending on what you're hearing. Something similar could occur in whales in a kind of tag-team hearing in which hemispheres coordinate processing based on which has the most refined circuits for handling a specific soundscape. How unihemispheric sleep affects the neural processing of echoes and the echolocation abilities of whales or dolphins remains completely unknown.

The observed similarities between plasticity-related circuits in whales and other mammals are encouraging. It suggests that studying the brains of smaller mammals with similarly structured brains may provide insights into what happens in a singing whale's brain. For instance, researchers recently measured the hearing abilities of the African pygmy hedgehog, an adorable pincushion that will fit in your hand and that is the model for all plush hedgehog toys. The cerebral cortex within a hedgehog's brain is organized more similarly to that of a humpback whale brain than is the case for most other small mammals.[26]

Hedgehogs don't sing, though some males do produce a birdlike sound in the presence of females that is called "the serenade." In principle, one could raise a pygmy hedgehog in a virtual world that simulates the auditory and visual experiences of a humpback whale[27] and then explore how its brain changes to accommodate this alien soundscape. Although the neural circuits formed by this humphog (hedgehump?) will differ from those of an actual whale, they could provide new insights into how a humpback's brain changes over time when the most important sounds it hears are constantly morphing vocalizations.

Studies with animal models can not only increase neuroscientists' understanding of how whale brains might work but may also lead to opportunities for watching whale brains in action. For example, new technologies are under development that make it possible to record the activity in marine mammals' brains without the need to capture them or subject them to surgery. So far, these portable and submergible brain-recording devices have only been used on dolphins, seals, and sea lions—both captive and in the wild. But in principle they might also be used to monitor the brain of a singing whale or a whale within the hearing range of a singer.

---

26. Cetaceans are genetically closer to insectivores (including hedgehogs) than they are to most other small mammals. By some estimates the genetic distance between humpback whales and hedgehogs is about four times smaller than between sheep and humpbacks.

27. Adjusted to fit the hearing sensitivities of hedgehogs.

Whale Song Acoustics: Songs as Sonar

Music can evoke feelings, questions, awe, actions, chills, and inspiration. Some people have similar reactions to whale songs, which effectively makes them musical, whether that was the goal of the singers or not. Listening humans may use whale songs to achieve states of blissful relaxation. But few if any have ever used them as acoustic floodlights. When thinking of songs as a kind of long-range sonar, what you hear is *not* what they get.

If you listen to bats echolocating, you will hear very little. Not echoes, nor anything else. The sounds are simply too high-pitched for your bulky auditory system to deal with. If you listen to dolphins echolocating, you're likely to hear something creaky. But you'd only be hearing the fringes of what dolphins are producing, and you wouldn't detect any echoes that they're using. Even if you could hear dolphin-generated echoes, they would reveal little to you. You would hear clicks, not perceive objects at a certain distance from your head. Whale songs are a worst-case scenario because you can easily hear the sounds singers are making along with lots of details of those sounds. Unfortunately, you'll be oblivious to most of the echoes falling in the "silences" between those sounds and doubly oblivious to any changes in those echoes that are occurring over periods of minutes.

Human brains are structured to mostly ignore reverberation—treating it as a kind of background noise—and to ignore any acoustic changes that are occurring over periods longer than tens of seconds. As a result, human listeners tend to perceptually erase the potentially crucial silences within songs and focus instead on the oddities of units grouped in patterns. For a singer listening to returning echoes, these would be the least relevant aspects of songs to hear. If whales are like dolphins and bats, then their brains are actively suppressing reception of the sounds they make while singing—essentially erasing from their perception the units and phrases that human listeners find so intriguing. Furthermore, what singing humpbacks perceive while listening is unlikely to be echoes from units, just as you don't perceive flashes of photons hitting your retinas. Instead, you see objects and people within scenes. Similarly, a listening singer would likely perceive either whales within scenes or empty seascapes rather than the echoes generated by specific units.

How then might one describe the properties of echoic scenes generated by songs? This is impossible by simply recording the units a singer produces, because the same song produced in different contexts will generate different scenes with different sets of "objects" (mainly whales) within them. It's possible, however, to track the ways that whales sing in a manner that more clearly reveals the kinds of echoes they're likely to generate at different points within a song, as well as the general kinds of information they might be able to extract from those echoes. This approach has been used to understand why bats change their cries as they home in on targets or navigate around obstacles.

The approach involves simply showing the full range of acoustic changes that an animal makes while vocalizing. In the case of singing humpback whales, this can be achieved using *singerings*, images that retain most of the acoustic features produced by singers. Singerings highlight persistent streams of units and how those streams shift over time. Singerings don't reveal the all-important echoes, but they can reveal the strategies singers may use when constructing echoic scenes.

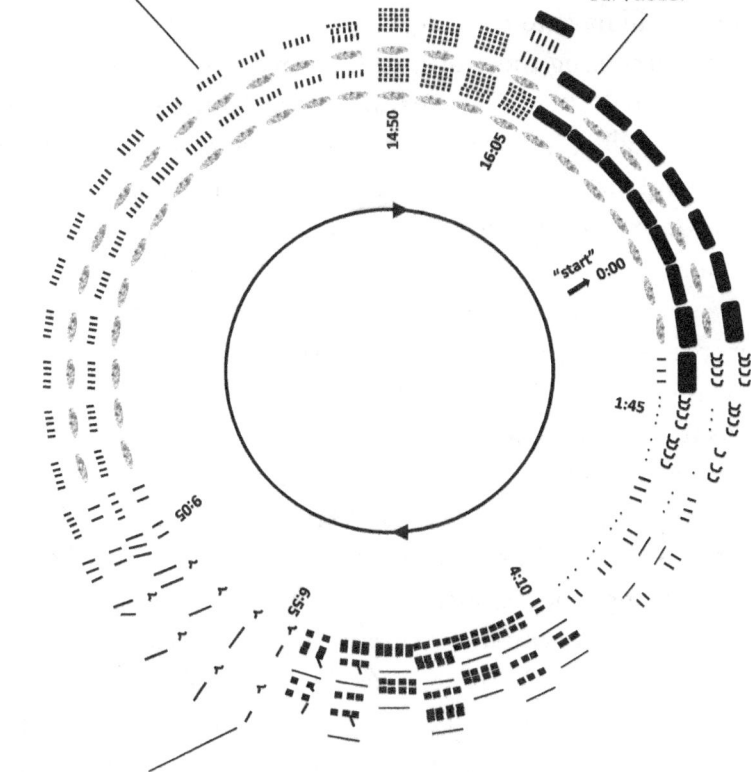

Concentric rings reveal regular rhythms.

Patterns like this show up just before a singer surfaces.

"start" 0:00

1:45

4:10

6:55

9:05

14:50

16:05

Theme transitions appear as shifts in patterns.

A singering is a circular arrangement of spectrograms of unit sequences within a song constructed such that consecutive song segments are aligned as closely as possible. The complete circle shows all the patterns a singer produced within a single song. Singerings can reveal the rate at which a singer gradually morphs unit features as well as the stability of unit timing across phrase repetitions, changes that affect the kinds of echoes that the song may generate as well as how those echoes might be processed by a singer.

The large heads of singing whales provide lots of spots where re-corders can be placed, and their large brains should generate rela-tively strong neural signals to record. Whales' massive head size also means, however, that any recording device placed on their head is going to have to contend with a thick skull and distant neu-rons, both of which will decrease the detectability of brain waves. Engineers will have to ramp up their creativity to figure out new ways of overcoming these obstacles before scientists will be able to get a good look at what's happening inside a singing whale's head.

Imagining what it's like for a singer to perceive distant whales using a potpourri of ever-changing echoes is a daunting task. Neuroscientific studies of whales and other animals may be able to establish what is physiologically possible, but they can't really clarify a whale's experience of returning echoes. A singing whale's experience may be much simpler than you might expect. Ulti-mately, the information relevant to an echolocating whale is the same as for a watchman or sentinel: Who is nearby and what are they doing? Perceptually, this is what matters. It's possible that all the complexity that scientists observe in songs is totally hid-den to the whales themselves. A singer's experience could simply involve noticing other whales moving about or waiting for whales to appear, without any awareness of units, phrases, themes, or songs or of the myriad ways that songs are morphing over time.

I've noticed over the last decade that some scientists are resis-tant to the sonar hypothesis not because of any specific counter-evidence but because they feel that this alternative is in some way a killjoy explanation that attempts to knock humpbacks off a priv-ileged pedestal. As long as humpback whale songs remain cultural traditions that males share in their attempts to impress discern-ing females, whales get to remain members of a special culture club that includes only a few cognitive elites, mainly primates and ce-taceans. This perspective is misguided because using sound as a form of active perception requires more cognitive sophistication than does reflexively regurgitating sound sequences that you hear around you. Echolocating singers have to not only transform sound waves into a dynamic social scene but also use that scene to predict the future: which whales will go where, how competitors will react to the situation, and what actions are required to end up in the right place at the right time. Simply copying sounds or actions requires no reasoning whatsoever, which is why describing someone's behavior as "aping" is an insult rather than a compliment. No matter how beautiful a peacock's tail is, no one's going to view those fancy feathers as evidence that peacocks are intellectual giants.

Singing whales have mammalian brains. They're likely pushing those brains to the limit when they sing, not because it's hard to sing an ornate song but because apprehending the world on a large scale without the benefit of vision is insanely difficult. This is something that humans can only pull off by using advanced technologies. Even with those technical aids, our capacity to perceive underwater scenes remains limited.[28] According to the sonar hypothesis, singing whales dynamically weave echoic soundscapes that they interpret in real time to make decisions about what to do next. They are constantly conscious oceanic nomads constructing a virtual world in their heads. It's a world that human activities are slowly dismantling, both by inundating singers with the sounds of ships and by erasing ancient circles of life through artificial global warming. Chapter 10 will consider the potential costs of these anthropogenic invasions for singing whales and for humans.

---

28. Think fish-finders.

CHAPTER 10

Will Whales Sing?

MY JOB AS A SCIENTIST is to identify viable hypotheses, assess them, and then refine or replace them when they falter. Well, that was my naive impression of the job description when I started grad school. It turns out that not all hypotheses are equal in the eyes of the experts—some carry dangers that others do not. Letting a woolly bear caterpillar explore your hand can be an endearing experience, the sort of science that children love. Not so with the Portuguese man-of-war,[1] despite their adorable balloon rafts. The hypothesis that male whales serenade females is a scientific woolly bear. Many a time I have thought to myself, "Put the man-of-war down." But, for better or worse, the "song as sonar" hypothesis has latched itself onto me. At the risk of exposing myself to further stings, here goes my best effort to recap the scientific support for the hypothesis that whales use songs for sonar.

The idea that humpback whale song might enable singers to see with sound was born from belugas, who sometimes group their sonar clicks into packets when inspecting targets from long distances. Effectively, belugas are "singing" with click packets, though

---

1. This oddity of nature, a colony of cloned zooids collectively masquerading as a jellyfish, packs a painful, poisonous sting.

no marine biologist would describe them that way. The lesson I learned early on from this cross-species comparison was that echolocation at long distances works a bit differently from short-range sonar. This was my first clue. Since that initial spark of possibility, further signs of sonar have come raining in.

First, let's consider the evolutionary context. Cetaceans gradually returned to their watery origins, where vision proved less useful. As has occurred in multiple species forced to make their way in the dark, sound became increasingly important as a source of information. Baleen whales clearly evolved in environmental conditions where the ability to use echoes would have been advantageous, and these abilities are known to have evolved in close relatives. There is every reason to believe that large whales would benefit from the ability to echoically perceive what is happening at long distances when that information is not accessible through any other means. Baleen whales did not evolve perceptual strategies that exactly match those that toothed whales evolved. But for them to not have evolved any means of using the echoic information produced by their sounds would be bizarre. Evolutionary theory cannot provide evidence that any behavior serves a specific function, but it can provide guidance on which capacities are likely to be favored in specific ecosystems. Other cetaceans clearly evolved sonar, so the onus is on scientists who think baleen whales missed the boat in this regard to explain why they failed to adapt.

The second, and in my opinion most important, clue to what's going on in songs comes from the repertoire of sounds used by singing humpback whales. It is the worst possible set of sounds one could imagine for communicating over long distances in shallow water. The sounds not only continually vary within songs but also vary across songs so that there is no fixed set of sounds like the vowels and consonants used by human singers. Add to that the highly distorting effects of sound transmission in shallow water and you end up with an unpredictable soup of sounds that other whales cannot reliably link to what the singer attempted to produce. Only the singer knows what a song was truly like when it emerged from its head. And for the singer, those same acoustic variations that complicate long-range communication can help determine the locations and movements of faraway targets. The choice is clear: Either one accepts that humpback whales evolved a horrifically screwed-up long-range communication system (and made their mating system dependent on it), or one entertains the possibility that they evolved a highly refined sonar capacity that they use to monitor the distant actions of others.

The evidence that convinced Donald Griffin and Ken Norris that bats and dolphins use echolocation came not from the sounds their subjects produced but from the way these animals behaved when navigating without sight. My initial reaction to the sonar hypothesis was "Surely someone would have seen singers approaching silent whales if they were using songs as sonar, and since no one has, the hypothesis is a nonstarter." Or had they? When I went back through early papers, there were reports of singers going silent and immediately rushing off to intercept other distant whales, as well as reports of whales engaging in evasive maneuvers to avoid being joined by a previously singing whale. These reports suggest that at least some singers are searching for and monitoring the movements of other whales. There really is no other way for singers to detect or locate these distant whales other than acoustically. The only question is this: Are distant whales only perceivable when they actively generate sounds (by vocalizing or slapping the water), or might they be detected based on the echoes they reflect? If whale songs generate detectable echoes from other whales, then this could explain how singers find, track, and join distant, silent whales.

Now, there is no question that sounds bounce off whales. Nothing that big with that much air in it is going to be transparent to sound.[2] Physics dictates that whales will reflect echoes. But which sounds will bounce back? And how far can they travel before they become indistinguishable from background noise? This is where scientists disagree. Simulations give widely varying estimates about what might happen. It's extremely tricky to record all the sounds received by a singer while tracking all surrounding whales that might be generating echoes. How detectable the echoes are will depend on lots of factors that are constantly changing, like the orientation, depth, and location of the whales. The echoes generated by song sounds will differ quite a bit from those perceived by echolocating dolphins or bats and are unlikely to be salient to human observers. What we do know is that songs generate strong echoes from landmasses and produce lots of long-lasting reverberation that is tightly focused at specific frequencies, meaning that when song sounds do generate echoes, those echoes propagate well for long distances. Even in the best-case scenario, however, echoes from distant whales will be tricky to detect and interpret, requiring some serious auditory processing by a singer's brain. That is what led me to consider exactly what whales' brains might be capable of when it comes to processing echoes and to the discovery that the humpback whale ear is off the charts with respect to how it neurally processes incoming sounds.

The acoustic fovea possessed by some species of bats is widely regarded as a specialization to enable the high resolution of behaviorally relevant echoic features.[3] As noted in chapter 7, the physiology of a humpback's ear suggests that a much greater expanse of its cochlea acts as an acoustic fovea than is the case in any echolocating bat. The only other mammals with anything comparable to the innervation present in baleen whale ears are toothed

---

2. Humpback whales show up really clearly on commercial fish-finders.

3. The nonacoustic fovea, referred to as "the fovea," is what enables vertebrates to finely resolve visual inputs, like you are doing right now in reading this sentence.

whales. No birds come close. The sonar hypothesis specifically predicts that the precise processing of incoming echoes is necessary for whales to track whale-sized targets at long ranges. So evidence of expanded frequency processing in a whale's ear provides strong support for this hypothesis.

Figuring out what singing humpback whales are doing with songs is not just an esoteric battle between whale nerds. Knowing how larger whales are using sound is important for predicting how human activities could affect them. Specifically, if singing humpback whales are tracking variations in faint echoes over time, then threats like ship traffic and climate change may seriously impact the ranges at which songs can function.

Human-Whale-Science Relationships: It's Complicated

In the 1970s, environmentalists compelled governments to protect whales from companies that were hunting them toward extinction.[4] Realistically, we don't know what whales' lives once were or what remains of their past life styles post whaling. The whaling industry targeted the boldest and most curious individuals, removing them from the gene pool. It's unclear how this artificial selection affected the diversity of current populations.

It's also not clear how effectively the Endangered Species Act (established in 1973) or the Marine Mammal Protection Act (established in 1972) protects large whales. Most of the human activities that put whales in danger today—oil spills, shipping traffic, military exercises, and climate change—are not curtailed

---

4. Part of the spark that led to increased public concern about the plight of large whales (and their smaller cousins) were the recordings of singing humpback whales popularized by Roger Payne. Public outcry was why (some) whales were "saved," but public demand for stuff was the main reason they needed saving.

at all by these laws. Ironically, the Marine Mammal Protection Act is more likely to hamper scientific studies of the threats facing whales than it is to neutralize those threats because scientists are required to obtain federal permits before they can legally conduct such studies. The process of obtaining a scientific permit to study large whales can take years and requires a significant amount of recordkeeping and reporting once obtained. In contrast, plowing into a whale with a large ship is relatively free and easy. There are rules about how fast ships can travel in certain areas inhabited by endangered species but no legal consequences when a boat kills whales by running them over. Many times, the captain never even knows what that bump in the night was.

One could imagine that, over millions of years, whales evolved mechanisms to face some life-threatening challenges. But the genetic specializations that provided resilience in the past could easily have overlapped with those weeded out by whalers. And many of the ecological challenges that whales face now have arisen recently, in a blink of evolutionary time.

Consider the potential effects of shipping traffic. Consumers' insatiable desire for new and better stuff necessitates that giant ships deliver the goods. These transoceanic delivery services account for nearly 10% of carbon dioxide emissions annually. As you might suspect, humongous boats do not tread lightly. They generate an impressive amount of noise, often in areas where humpback whales sing (or used to sing). Shipping noise dramatically shrinks how far whale sounds can travel, making it harder to be heard. In some cases, whales persevere and continue singing until a ship

Overleaf: *When worlds collide.*
Large cargo ships often move through areas where singing whales are (or were) present. The loud noises generated by these ships obliterate whales' songs.

has passed. It's not rare, however, for multiple ships to pass by back-to-back, in which case singers often give up. We don't know whether such experiences affect where and when whales choose to sing or how many options singers have when it comes to choosing a good singing spot. Without knowing what whales are attempting to do when singing, it's hard to predict the conditions or locations necessary for their songs to work well.

In recent years, scientists have ramped up their efforts to understand how changing ecosystems may affect cetaceans' ability to use sound. Scientific conferences now devote entire sessions to this topic. In the case of singing humpback whales, scientists have mainly focused on whether increased noise from ships causes whales to leave, stop singing, or change the way they sing. All these outcomes have been documented, but the consequences are unclear. After all, singing whales move to different locations, stop singing, and change the way they sing all the time. Does it really matter if they do it a bit more? Perhaps for whales, being exposed to shipping noise is similar to when you wave your hand to discourage an insect from buzzing around your head—an annoying incident but one that has no major implications for your life goals.

Increased shipping is just one of the ways that modern humans are blindly tinkering with whales' singing behavior. Industrially driven increases in ocean temperatures can have much broader effects, both directly and indirectly. I noted in chapter 4 that how songs (and the echoes from songs) propagate in water depends partly on temperature variations. Changing temperature profiles in habitats where whales currently sing will change how their songs propagate. Maybe singers can adjust to account for those effects and maybe they can't. Increasing temperatures also strongly affect the distribution of prey that whales eat. Such changes will affect how much food is available, where food is available, and when it's available. What will that do to whales' migration patterns? And if whales start migrating differently, might that affect when and where they sing? Again, without knowing how whales are using songs, it's hard to say which of these newly imposed scenarios might impede functional singing.

If songs are acoustic peacock tails, then a little bit of ship traffic blaring overhead may be bearable—like living near an airport—and warmer temperatures should have little effect on songs. Most scientists currently assume that singing is one of multiple mating strategies, so if songs stop working, male humpback whales can always fall back on the old-fashioned approach of beating each other up until the winner claims the spoils. If songs are how singers perceive their world, however, it's a totally different story. Because if whales lose the capacity to track each other's actions, then they may not have a backup strategy, like drivers attempting to navigate in thick fog or during a blizzard.

Changes in water temperature can have large effects on how echoes travel underwater, and ship noise can make the detection of faint echoes much more difficult. If humans persist in "dimming the lights" on singing whales, we could silence them permanently without ever fully understanding their songs. Our ignorance may finish the job whalers started. Whalers accidentally changed the genetic profiles of whales by "harvesting" those that yielded the greatest profit. Similarly, Amazon Prime members may inadvertently be setting the stage for an even more devastating cull.

What Would Be Lost?

Although humpback whale numbers have grown over the past couple of decades, this recovery is precarious at best. What will be lost if extinction silences the songs of whales? The implications go beyond a decrease in biodiversity and the death of the whale-watching industry. We could potentially lose an intelligent mind with unique ways of experiencing the world.

Archeological studies of Neanderthals provide insights into what will be missing if singing whales are no more. You have never met a Neanderthal, but if you did, she would be recognizably human. Archeologists used to depict them as Quasimodean humans, but recent discoveries suggest this caricature was a mistake. There is now evidence of Neanderthal art, music, medicine, cooking, tool construction, and seafaring. Their brains were larger than yours. Maybe they had spoken languages, but there is no way to know for sure. Only hints of what Neanderthals felt and thought remain in the traces they left behind. Neanderthal minds are permanently lost. Whatever they might have revealed to us about ourselves or the nature of selves and minds, we will never know.

Unlike Neanderthals, whales leave no durable signs of minds; indeed, it's not clear to everyone that the term "mind" even applies to whales. What do whales think and feel? Are they aware in a way that we as humans would understand? It's now firmly established that animals can perceive energetic events that humans have no natural ability to detect. For instance, sharks perceive magnetic fields, platypuses sense electric pulses, and snakes form thermal images from infrared radiation. Nevertheless, people still cling to the notion that when it comes to emotions, cognition, and conscious states, we are as advanced as it gets. We tend to think that if other species were operating at a higher level than humans, then they would better protect themselves or would create artifacts more impressive than those made by humans. This dismissive tendency is a prime example of *anthropocentrism*, the viewpoint that modern adult humans provide the gold standard for what is mentally possible and exceptional. From an anthropocentric perspective, "higher-level" mentality basically means whatever an impressive adult human can do plus a bit extra.

The first scientific book to address issues of animal awareness was written by Donald Griffin of bat sonar fame. The book has little to say about what whale minds are like, but Griffin does argue that "the better an animal understands its physical, biological, and social environment, the better it can adjust its behavior to accomplish whatever goals may be important in its life, including those that contribute to evolutionary fitness."[5] Essentially, he suggests that awareness is adaptive. Many others have picked up on this line of argument, including most recently Peter Godfrey-Smith, who suggests that monitoring sensations generated by actions was the key step leading to the evolution of awareness. If active engagement with the world is what led to the evolution of minds, then there is every reason to expect that the minds of whales with the capacity to echolocate over vast expanses would be exceptionally aware.

A singing whale's awareness of itself and others may gradually arise from its ability to notice characteristics of echoes that initially were too faint to be distinguished from background noise. From this perspective, singers' mental states are determined by their ability to make subtle perceptual distinctions—if you cannot discriminate red from green or brown, you cannot experience "redness." To the extent that singing whales can make distinctions that you cannot, their levels of awareness may exceed yours. Just as infants are unable to discern that their parents are aware of things beyond their understanding, adult humans may fail to recognize when another species is aware in ways that extend beyond what is normally possible for humans.

Why do whales sing? The "song as sonar" hypothesis says that humpback whales (and probably other singing whales) produce sounds hours on end to perceive the world around them, particularly to explore unpredictable scenes. They sing to actively form soundscapes. Through vigilant echolocation, singers can increase their chances of being in the know. A silent whale is entirely dependent on others to reveal what is happening where. Singing

_____

5. From *The Question of Animal Awareness*, page 145.

whales create awareness, enlightening themselves and others. But what exactly is their constructed awareness like? Is it like a guard controlling a searchlight? Or a toddler feeling for a ball that has rolled beneath a sofa? Or is a singing whale's experience radically different from anything a human has ever perceived? Humans can eavesdrop on whales' craftmanship, but they are unlikely to share the awareness that whales gain from songs. We are simply not equipped to apprehend complex echoic panoramas. It would be a shame to lose the few minds that do possess this capacity due to a seemingly perpetual obliviousness about how our actions impact marine environments.

That a singer's awareness of ongoing events is likely to differ radically from that of a human becomes particularly clear when one considers more closely the typical conditions within which singing occurs. *** WARNING: If you suffer from thalassaphobia, skip to the next paragraph. *** Imagine you are out in the open ocean, perhaps a mile from shore, at night, by yourself, submerged and holding your breath. Now imagine you maintain this position, alert, for 8 hours (with a short break to breathe once every 15 minutes). Many people would view spending even 5 seconds in this scenario as a good reason to panic.[6] Carl Jung included this kind of aquatic fear in his archetypes of the collective unconscious—universal symbols that shape how humans view the world and that lead to shared myths and legends across cultures. Clinicians now refer to this fear as thalassaphobia, a persistent and intense fear of deep, dark water. The subgenre in which one is specifically terrified of unseen, many-toothed lurkers below is dubbed megalohydrothalassaphobia.[7] It seems certain that few if any whales suffer from either of these conditions. They

---

6. Did I mention how there is no way to know what's lurking beneath you?

7. The fear of jumping off a diving board into the ocean at night because of hidden creatures below is catapedamegalohydrothalassophobia.

may, however, have an intense fear of lying on a beach for several hours as the sun passes slowly overhead. Your nightmare is their singing sweet spot, and their daymare is your vacation spot.

Eastern philosophies may offer some clues about the levels of awareness that singing whales naturally achieve. The Buddha described a set of *jhanas*, or meditative practices, that could be used to achieve a state of nirvana in which one's mind is "concentrated, purified, bright, unblemished, rid of imperfection, malleable, wieldy, steady and attained to imperturbability."[8] The first jhana requires a withdrawal from sensuality, restlessness, doubt, ill will, and sloth. The lone singer, drifting suspended beneath the surface, seems likely to be free from these sorts of distractions. The second jhana involves freeing the mind of all thoughts and achieving a state of unified and one-pointed awareness. A singer focusing all its attention hour after hour on interpreting the surrounding echoscape will necessarily achieve such a state.[9] In the third jhana, the practitioner focuses on mindfulness, sameness, and discernment. An echolocating singer must constantly evaluate incoming echo streams to discern signs of other whales within its range of surveillance. The fourth jhana is a final state of unwavering concentration in which happiness is abandoned and replaced with a sense of stability. Whether singing whales ever achieve this fourth state is debatable, but if they are happy while singing, then they are good at hiding it. Humpback whales' concentration while singing appears as unwavering as that of any Buddhist monk.

What is the cost of a lost mind? Buddhists, who view the physical dissolution of minds as a natural transition, might say that the question is meaningless. What is the cost of a lost song? Some may mourn the haunting melodies flowing through the ocean's

---

8. From the *Visuddhimagga* (The Path of Purification), a foundational text in Theravāda Buddhism written by the scholar Buddhaghosa.

9. This state may also be akin to the psychological state of "flow."

vastness, those intricate, visceral vocalizations that remind one of the mysterious interconnectedness of all life on this planet. Certainly, there are those who cannot bear the thought that the musical creations of these charismatic creatures might soon be no more. But what if they never were? Songs, that is. What if the "songs" of whales are simply a romantic notion that has run its course? Do whales need to sing to warrant our consideration?

If humpback whales had not been characterized (caricatured?) as singers, then they likely would have met the same fate as the passenger pigeon—a species wiped out through overhunting and habitat destruction.[10] Whales would not exist and neither would this book. Science is often sold and extolled as a path toward enlightenment, a way to solve problems, advance knowledge, increase understanding, and explain the nature of the world. A less appreciated virtue of science is its capacity to correct mistakes. It wasn't a mistake to deem whales singers or to promote their vocal skills to save them. The proposal that humpbacks sing like birds propelled scientific studies of cetaceans and helped save them from extinction. It is a mistake, however, to assume that whales will sing on as long as humans don't hunt them. We now have the means to reveal cetaceans' capacity to perceive sonic scenes. Only by delving into the depths of whales' true experiences can we hope to understand the potential consequences of our ongoing actions. For their sake and ours.

10. There were more than 3 billion passenger pigeons living in North America in the late 1800s, when people began killing them in the millions.

Epilogue

I don't want to leave you with the misimpression that my early ideas about the nature of whale song, disparaged in my days as a grad student, have risen from the ashes to revolutionize cetologists' views on why whales sing. On the contrary, the sonar hypothesis continues to be reviled and ridiculed by many marine mammal scientists. Just last week, I received a scathing review of a scientific paper I submitted describing a surprising acoustic property of humpback whale songs. In a sentence near the end of that paper, I noted in passing that the observed song feature could affect how precisely a singer would be able to localize other singers or echoes from other whales. This apparently triggered one expert reviewer, who, after informing the editor that the paper had no theoretical justification, questionable methods, and "some vague and poorly specified arguments," went on to say: "There is a half-hearted attempt to revive a theory about echolocation which makes no sense at all (why only males? why uniquely among echolocating animals would humpbacks have such a hugely variable signal—no other echolocating animal varies their signals in anything like the dramatic manner humpbacks do). This fringe idea has been long debunked (Au et al 2001) and persists only in zombie form." Readers beware! You're at risk of being bitten! This, sadly, is often the way science works. Nobel Prize–winning physicist Max Planck famously observed in his autobiography that "an important scientific innovation rarely makes its way by gradually winning over and converting its opponents: it rarely happens that Saul becomes Paul. What does happen is that its opponents gradually die out."

Current AI platforms have a more nuanced view on the singing behavior of whales than do most whale researchers. When asked why whales sound like singers, ChatGPT 3.5 replied, "Some whales sound like they are singing because they produce complex patterns of vocalizations, possibly for communication, social interaction, navigation, or other purposes, which, to human ears, resemble melodic singing." Not only did this artificial neural network present a range of possible functions for whale songs, including sonar (within the category "other purposes"), it also pointed out that human perceptual categories ultimately determine which phenomena observers count as "singing."

While it's true that researchers studying animal vocal behavior have provided objective criteria for deciding if an animal's vocal act should be described as singing versus calling, those criteria have not been consistently applied. When zoologist David Spector reviewed 80 scientific definitions of the term "bird song," he found that there was little agreement about what defines a song or what differentiates songs from calls. He also noted that definitions changed over time, and not in ways that made them more definitive. Based on his findings, he concluded that a song definition that covers all concepts of song is not possible.

In the case of cetacean vocalizations, use of the term "song" has clearly done good, both in terms of engaging the public to consider whales' lives more compassionately and in terms of inspiring scientists to study their cognitive and social capacities. On the other hand, early adoption of this descriptor may have channeled scientific thinking about cetaceans in ways that have hampered advances in understanding the vocal actions of both whales and dolphins.

Based on the criteria that Roger Payne and Scott McVay set down in their seminal *Science* paper describing humpback whale sound sequences as song, dolphins also "sing." Specifically, various toothed whales produce "series of notes, generally of more than one type" in ways that form "a recognizable sequence or pattern in time." Some of those patterns (such as call sequences produced by orcas) are more structurally and acoustically complex than the "songs" produced by blue and fin whales. Scientifically, it would be as justified to claim that dolphins sing and baleen whales don't as to claim the reverse.

It's relatively easy to avoid such definitional hypocrisies. For instance, one could say that cetaceans produce vocal sequences in recognizable patterns. If this seems like a less provocative or compelling way of describing what a particular species does vocally, then that's more an issue of personal aesthetics than science. As a singer who has performed in a variety of choral contexts, I appreciate that describing whales as singers fosters a kind of cross-species comradery that describing them as sound producers or vocalizers does not. When it comes to deciding how best to comprehend cetaceans' actions and minds, however, there are advantages to maintaining some aloofness.

As listeners, humans have no choice but to hear the songs of whales as a series of sounds. Many of the patterns within those sequences are not naturally perceptible—that's why engineers listening to humpback whales didn't immediately realize that there were multiminute long cycles in the sequences they were hearing. Similarly, no human is going to naturally focus on faint echoes within the intervals between units, much less transform those echoes into a three-dimensional spatial landscape revealing the positions of silent animals. At present, we don't really know how whales perceive their songs. Where humans hear songs, whales may construct worlds. Respecting such differences in perspective may be key to clarifying what the lives and experiences of both whales and dolphins are truly like.

Regardless of whether you believe dolphins or whales sing to serenade others or to enlighten themselves (or both), I hope this book has helped you to imagine the lives of other species in new ways and to better understand the forces that shape how scientists explore and interpret the world.

ACKNOWLEDGMENTS

The ideas within this book emerged over a 30-year span. My delible memory precludes the possibility of me thanking all the people who have shaped my thinking about whale perception and cognition across 3 decades. So if you're not noted below and feel I've failed to recognize your critical contributions, please feel free to find me and poke me with a pencil.

In the beginning, there were books. Books that sparked the seed of a thought that maybe, possibly, I could conduct studies of cetacean minds. Books by John Lilly, Herb Roitblat, Lou Herman, Whit Au, and many others that convinced me that the mental capacities of dolphins could be scientifically studied. I was lucky enough to be able to bounce my ideas about cetacean vocalizations off all these authors as a grad student in Hawai'i. The scientists who ultimately guided my thinking about whale songs, however, were not whale researchers but ocean acousticians. Neil Frazer and Alex Tolstoy enlightened me on the complexities of shallow-water propagation and made it possible for me to simulate long-range song transmission as opposed to just imagining it. Neil is the reason this book exists because he was the only scientist who encouraged me to explore a novel hypothesis that everyone else deemed ludicrous.

Deirdre Killebrew inspired me throughout my doctoral research, helping me survive three theses as well as the complexities of cetacean research at the Kewalo Basin Marine Mammal Laboratory. Scott Murray and Carrie DeLong were beacons of brilliance in an ocean of marine mammal enthusiasts. They greatly helped me refine my thoughts about cetacean cognition. I've supervised three doctoral students interested in knowing more about singing humpback whales: Jennifer Schneider, Sean Green, and Christina Perazio. Although I did my best to shield them from the stigma of contemplating the possibility that singing humpback whales might be echolocating, their persistence in pursuing research on singing whales is largely what kept me from abandoning the field.

Although the ideas within this book were born in the 1990s, the book itself did not begin to take form until 2010 during my time as a fellow at the Center for Advanced Study in the Behavioral Sciences at Stanford. Most of the book was written in 2023 and 2024, through the support of the Alfred P. Sloan Foundation, the John S. Guggenheim Foundation, and the Harvard Radcliffe Institute. As a Radcliffe Fellow, I was privileged to be part of an interdisciplinary writing group of fellow fellows. I thank Laura DeMarco, Peter Gray, Ruth Grossman, Judith Lok, Narges Mahyar, Valentina Rozas-Krause, and Kim Vaz-Deville for their participation in that group and for their weekly encouragement, which helped me to steadily progress in writing chapters.

I'm particularly grateful for the help and guidance provided by Tiffany Gasbarrini, my editor, toward making this book a reality. Mary Bates was the best possible developmental editor I could have hoped for. She helped to shape every coherent sentence in this book. I thank the talented Emily Eng for her careful crafting of the underwater scenes featured throughout the book and for patiently modifying these images to capture the odd and obscure elements I envisioned. I also want to thank the rest of the editing and design team—Ezra Rodriguez and Molly Seamans at Johns

Hopkins University Press and Helen Wheeler at Westchester Publishing Services—for launching *Why Whales Sing* into the world.

A special thanks goes to Brian Branstetter, who agreed to serve as an unofficial spokesperson for the field of cetacean bioacoustics. Though I've failed to sway him to my way of thinking when it comes to the songs of whales, I'm glad he does not feel the compulsion, common among some in the field, to call me a dumb ass, despite his skepticism.

I'm grateful to my mom, Donna Shaw, for being supportive of my scientific efforts throughout the past 30-plus years and to my wife, Itzel Orduña, for putting up with my incessant typing and staring at images of spectrograms for nearly as long.

Finally, I thank Ava Brent for motivating me to persevere in writing this book, long after I had given up on the project as a lost cause. Her passion for science, whales, and scientific stories about whales provided the catalyst I needed to make one last attempt to convince a publisher that people might be interested in knowing more about the mental lives of whales.

APPENDIX: ESSENTIALS
OF ECHOLOCATION

If you've ever gone camping or experienced a blackout, then you're familiar with the challenges of finding your way in darkness. Similar conditions led to the evolution of echolocation abilities in bats and cetaceans. Don Griffin introduced the term *echolocation* in the 1940s to describe a process of locating obstacles by listening for echoes. This ability, also referred to as animal sonar or biosonar, goes far beyond locating and avoiding obstacles, however. Perceptually, echolocation is thought to be like a combination of X-ray vision, hearing, and touch. Like most mammals, echolocators can tell where sounds are coming from by listening. Unlike most mammals, they can also sense the sizes, shapes, and detailed movements of objects from hearing faint echoes.

How Do They Do It?

Echolocation is often compared to using a flashlight because it reveals scenes that would otherwise be hidden and because bats and dolphins actively point their voices toward the locations they are investigating. Unlike flashlight wielders, however, echolocators aren't continuously spewing out energetic waves to guide their movements. Dolphins echolocate to expand their experiences, much like a child might reach out to touch a kitten. Although you can know a lot about a kitten just by looking at it, even young children know that you can't fully experience a kitten unless you're cuddling it. Dolphins and bats don't need to echolocate to successfully navigate—they often rely on vision, silent listening, touch, and other sensory modalities to guide their movements. But echolocation can make animals aware of details they wouldn't know about otherwise.

Fundamentally, echolocation is a mode of hearing. But perceptually, echolocation depends less on *what* sounds are heard and more on *how* sounds are heard. Although you can hear some of the sounds that dolphins produce when echolocating, you can't easily navigate or recognize objects using the echoes those sounds generate. That's partly because humans can't hear the high pitches that dolphins and bats hear and partly because humans' brains didn't evolve to interpret faint echoes as features of small objects.

At the most basic level, animals echolocate by comparing the echoes they hear to the sounds that generated the echoes. The echoes are like quieter replicas of the original sound, just like the kinds of echoes you might make by yelling at a deep ravine. The main difference is that when you hear an echo of your yelled "Hello!," you perceive the echo as if someone else had yelled hello from far away, while echolocators perceive reflected sound waves as friends, foes, food, or scenery.

Echolocation is particularly useful for determining the distance to objects. Echolocating bats and dolphins can accomplish this by neurally computing the amount of time it takes for an echo to reach them: Echoes that return faster indicate shorter distances. This timing needs to be precise and quick, even faster than your perception of a flash of lightning. Many of the specializations seen in the auditory systems of bats and dolphins are adaptations for rapidly processing incoming echoes.

Simply knowing how far away something is does not tell you where it is. You also need to know a direction. Is it above or below?

In front or behind? To the left or the right? Echolocating bats and dolphins mainly focus on perceiving echoes coming from the direction they are facing. Bats' focal field of "view" tends to be broader than that of dolphins. But both need to turn their heads and bodies to sense objects that aren't dead ahead. Bats use subtle changes in echoes produced by their head and ears, along with differences in timing, to figure out what direction echoes are coming from. It's less clear how dolphins manage this feat because they have no external ears, and sound travels from water into their heads almost as if their heads were absent. It's generally thought that internal structures within a dolphin's head affect incoming sounds in ways that are something like ear flaps. But no one is certain what cetaceans' "internal ear funnels" look like.

Ultimately, the properties of echoes that an echolocator senses at each of its two ears determine which features of objects it perceives. For instance, an echolocating bat can tell whether an escaping beetle is flying upward or downward based on subtle differences in how pitches change in each ear. The differences between successive echoes are probably more important than each individual echo in determining what an echolocating bat or dolphin perceives. It's similar to how the integration of visual information over time and across your two eyes largely determines what you see.

Unveiling a World of Reflections

Before the discovery of echolocation, scientists argued that bats' agile night flights were guided by their "sensitive wings." The idea was that bats quickly felt obstacles and avoided them. Or bats felt changes in air currents caused by solid objects, thereby detecting them. This explanation dominated until the early 1900s despite demonstrations that if you coated a bat's wings with three coats of varnish it had no effect on their ability to avoid objects and despite evidence that if you plugged bats' ears they started running into obstacles. Why did most scientists ignore or never hear about this damning counterevidence? Because a highly respected scientist, Georges Cuvier (1769–1832), dismissed the findings as flawed.[1]

Surprisingly, it was a curious undergraduate who made the breakthrough observation that bats were producing ultrasonic calls while flying and who recognized that these inaudible calls could explain the apparent clumsiness of deafened bats. In the late 1930s, Don Griffin convinced a Harvard physicist working on a fancy speaker that made ultrasonic sounds audible to try it out near bats. Once scientists were able to directly perceive the sounds flying bats made, they rapidly established that bats were using ultrasonic echoes to avoid obstacles and catch prey.[2] Despite clear behavioral, acoustic, and physiological evidence that bats were producing and perceiving ultrasonic calls, Don's suggestion that bats may use echoes to navigate was initially met with skepticism and viewed by some scientists as crazy. He was only able to convince others of his findings and interpretations by recording films of his experiments and by inviting skeptics to view live demonstrations of the echolocating bats in action.

Another two decades passed before experimenters established that dolphins also echolocate with ultrasonic vocalizations. This was shown by blindfolding captive dolphins and having them swim through obstacles while being recorded. Unlike with bats, researchers studying cetaceans were aware that dolphins produced rapid trains of clicks long before they realized dolphins used those clicks to echolocate. That's because humans can hear the lowest pitches

---

1. He also claimed there was no evidence for evolution and strong evidence for racial differences in mental abilities.

2. Even Don was dubious at first. In his initial paper describing bats' ultrasonic calls (published with G. W. Pierce in the *Journal of Mammalogy* in 1938), he concludes on page 455 that "the absence of the high frequency sounds when the bat was approaching obstacles suggests that they represent a call note or alarm signal of some sort rather than a means of avoiding obstacles." Only later did he realize that the calls were highly directional so they could not be detected from behind the bats.

present within the clicks. Before the discovery of bat echolocation, most people assumed that dolphin clicks were like the grunts of pigs, mainly serving as social signals.

Echolocation was initially shown in one species of bat (the big brown bat) and in one cetacean (the bottlenose dolphin). Following the discovery of echolocation in these two species, investigators began searching for evidence of this ability in many other animals. These follow-up studies revealed that different species echolocate in different ways. The specific sounds and echoes used by each species reflect the habitats within which they echolocate and the goals they pursue while echolocating.

Different Echoes Provide Different Information

Every sound can potentially bounce off a surface, but not all sounds rebound in the same way. Physics makes it possible to estimate how effectively a specific sound generates echoes when it hits surfaces of various sizes. To keep things simple, imagine that there are only two kinds of sounds: sounds that are long and whistly (think flute) and sounds that are short and explosive (think drum). The kinds of echoes that each can generate differ considerably. As a result, the kinds of information that an echolocating bat could get from hearing echoes produced by an ultrasonic drum or whistle are not the same. Echoes from drum-like sounds are short, making it easier to precisely time when they arrive. Echoes from whistly sounds are not so easy to time. Whistles tend to travel longer distances, however, so they can generate detectable echoes from things that are farther away.

These physical constraints have shaped the evolution of echolocation in both bats and dolphins. For instance, when some bats are searching for flying insects from long distances, they tend to produce calls that are more whistle-like. In contrast, when they are close to gobbling up a meal, these same bats produce very short calls that sound more like a drumroll. The bats essentially switch between the calls they use the way golfers switch between different clubs as they get closer to a hole. When trying to generate echoes from long distances, echolocating bats use calls with more oomph so they can "hear far." Once bats are close to the goal, they switch to using sounds that make the timing of echo arrivals more obvious. More precise timing means more clues about where a target is and where it's going.

Different species of bats live in different habitats and specialize in exploiting different food sources. Consequently, each species has evolved somewhat different echolocation calls and strategies. If you could hear echolocating bats, you could learn to distinguish different species based on what they sound like, much like singing birds. Bats that show the most flexibility in terms of how and what they hunt also possess the most varied vocal repertoires.

The links between bats' echolocation calls and their lifestyles is so tight that scientists can predict a species' ecological circumstance from the calls it uses. For instance, a bat that produces longer-duration whistly calls at a relatively slow pace (leaving longer pauses between calls) is likely to be focusing on finding larger, slower-moving food from relatively long distances, in a habitat where there are few obstacles.

Most of what is known about cetacean echolocation comes from studies of bottlenose dolphins. The clicks that bottlenose dolphins use to echolocate are much shorter than the calls of echolocating bats and much less variable. This is probably because sound travels five times faster in water than in air, meaning that sounds need to be much shorter for a listener to be able to time differences in when echoes are arriving. Click-based echolocation underwater also tends to provide a more spatially limited view than those experienced by echolocating bats, making dolphins' perception of echoes more like viewing the world through binoculars. On the plus side, sounds travel more effectively in water so that dolphins can detect echoes from distances 10 times farther than what bats can manage.

Echoes received by bats in air are mainly caused by surface reflections. In contrast, the strongest echoes received by dolphins and other cetaceans come from air pockets. This means that when a dolphin echolocates another dolphin, the strongest echoes will come from internal cavities (especially the lungs), the next-strongest echoes will come from dense bones (like the skull), and the weakest echoes will come from those fleshy, fishlike features that fascinate humans. Perceiving your friends and family as skulls levitating over pairs of glowing lungs may sound like nightmare material to you, but it's daily life for a dolphin.

Although echolocating dolphins do not gradually morph the duration or pitches of their clicks as they home in on prey in the way that bats do, they do vary the timing of their clicks. As a result, you can tell how far away a dolphin is from something it's echoically investigating by listening to how rapidly it's producing clicks—faster means closer. When dolphins are scanning targets that are far, far away, they often switch to producing clicks in short bursts. Presumably, this enhances their ability to detect very faint echoes, but exactly how bursting clicks helps with long-distance echo detection is not known.

Bottlenose dolphins echolocate using percussive clicks. Other toothed whales, like harbor porpoises, use more whistly sounds. As with bats, different cetacean species likely evolved echolocation calls that are customized for their specific ecological conditions. Much less is known about the perceptual circumstances faced by different species of cetaceans than is the case for bats, however, making it trickier to figure out exactly what benefits their vocal variations provide.

Although there is no way to be certain, it seems likely that variations in echolocation calls enable echolocators to search for targets in selected locations and to focus on specific details of detected targets. Call selection and adjustment would then be analogous to how you shift your gaze to peer at objects that catch your interest—if you varied your gaze based on where you were. Echolocators can vary the duration, intensity, pitch content, and form of their calls in ways that dynamically affect their ability to resolve specific objects (and their actions) at varying distances.

Despite vast differences in the circumstances faced by swooping bats and diving dolphins, the ways they use echoes to find and track prey are not so different. This convergence probably occurred because both groups mainly use echolocation to monitor rapid changes in their surroundings and details of scenes they cannot obtain through vision alone. Short-duration, ultrasonic vocalizations produced in rapid trains are ideal for the precise strikes that dolphins and bats make while intercepting and capturing prey fleeing in three dimensions.

Higher-pitched sounds produce stronger echoes from tiny targets and can more clearly reveal the direction from which echoes originate. A major disadvantage of ultrasonic echolocation, however, is that higher-pitched sounds will propagate shorter distances than lower-pitched sounds in both air and water. When an echolocator's goal is to intercept and capture small, fast-moving prey, the benefits of ultrasound outweigh its costs. If an echolocator wants to detect larger, slowly moving targets,[3] then lower-pitched and even audible sounds may be more effective.

What Do Echolocators Perceive?

When Don Griffin first coined the term echolocation, he envisioned an ability somewhat like using a cane to navigate a sidewalk. But in the decades since then, scientists have discovered echolocators can do so much more. Echolocation seems to yield perceptions analogous to those humans typically gain through

---

3. A distant tree would be a slow-moving target for a flying bat because although trees are stationary, the echoes they generate would be the same as if the bat is stationary and the trees are moving.

vision. It also provides information that goes beyond vision. For example, an echolocating dolphin can tell if a pregnant woman has twins by listening to echoes coming from within her stomach. Dolphins can also detect hollow chambers within metal balls, discriminate the metals used to make the balls, and detect such balls buried deep beneath rocks and sand (like a living metal detector). Recent studies have shown that dolphins can immediately recognize unfamiliar shapes previously viewed echoically the very first time they see them. In other words, dolphins can visually identify shapes based on past experiences hearing echoes from those shapes.

Experiments have also revealed bats to be sophisticated echolocators. For instance, bats can predict the future position of a moving target as it passes behind an occluding barrier. This means that bats are not only perceiving moving objects based on the echoes from those objects but also imagining where those objects will be in the future. Additionally, both bats and dolphins can make use of echoes produced by nearby comrades to recognize and track objects. This ability is comparable to you feeling the shapes of objects that a nearby friend is fondling.

For these reasons, some scientists now view the term echolocation as a misnomer. Echolocators are not simply localizing sounds. They're using reflected sounds in sophisticated ways that humans are just beginning to comprehend. Neuroscientists are increasingly able to track the neural activity produced when bats hear returning echoes. But they've yet to fully understand what echolocators perceive when they hear echoes. Bat and cetacean brains generate some sort of perceived scenes from detected echoes. Maybe their experiences are like those of seeing, hearing, or feeling. Or maybe they're unlike anything a human has ever experienced. Are echoic percepts blips like the images revealed by a strobe light? Or perhaps they're more like the stable visual scenes you see as your blinking eyes jerkily scan a room? When a bat or dolphin aims its acoustic gaze in your direction, your (and my) conception of the consequent perception is simply a provisional abstraction, a presumptuous predication awaiting transformation via further information. For now, it's time to say sayonara to sonar and to move toward new ways of solving the mysteries of leviathan minds.

FURTHER READING

Griffin, D. R. (1976). *The question of animal awareness: Evolutionary continuity of mental experiences*. Rockefeller University Press.

Herman, L. M. (Ed.) (1980). *Cetacean behavior: Mechanisms and functions*. Wiley Interscience.

Chapter 1. Why Whales Sing

Payne, R. S., & McVay, S. (1971). Songs of humpback whales. *Science, 173*, 585–597.

Rothenberg, D. (2008). *Thousand mile song: Whale music in a sea of sound*. Basic Books.

Winn, H. E., Perkins, P. J., & Poulter, T. C. (1970). *Sounds of the humpback whale*. Proceedings of the 7th Annual Conference on Biological Sonar and Diving Mammals, Menlo Park, CA, Stanford Research Institute.

Do Whales Sing?

Bradbury, J. W., & Vehrencamp, S. L. (2011). *Principles of animal communication*. Oxford University Press.

Fitch, W. T., & Jarvis, E. D. (2013). Birdsong and other animal models for human speech, song, and vocal learning. In M. A. Arbib (Ed.), *Language, music, and the brain: A mysterious relationship* (Vol. 10, pp. 499–539). MIT Press.

Frazer, L. N., & Mercado, E., III. (2000). A sonar model for humpback whale song. *IEEE Journal of Oceanic Engineering, 25*, 160–182.

Glockner, D. A. (1983). Determining the sex of humpback whales (*Megaptera novaeagliae*) in their natural environment. In R. Payne (Ed.), *Communication and behavior of whales* (pp. 447–464). Westview Press.

Herman, L. M. (2017). The multiple functions of male song within the humpback whale (*Megaptera novaeangliae*) mating system: Review, evaluation, and synthesis. *Biological Reviews, 92*, 1795–1818.

Moss, C. F., Chiu, C., & Moore, P. M. (2014). Analysis of natural scenes by echolocation in bats and dolphins. In A. Surlykke, P. E. Nachtigall, R. R. Fay, & A. N. Popper (Eds.), *Biosonar* (pp. 231–256). Springer.

Payne, R. (1995). *Among whales*. Scribner.

Tyack, P. (1981). Interactions between singing Hawaiian humpback whales and conspecifics nearby. *Behavioral Ecology and Sociobiology, 8*, 105–116.

Winn, H. E., & Winn, L. K. (1978). The song of the humpback whale *Megaptera novaeangliae* in the West Indies. *Marine Biology, 47*, 97–114.

Plastic Whales

Au, W. W. L. (1993). *The sonar of dolphins*. Springer Science & Business Media.

Helweg, D. A., Frankel, A. S., Mobley, J., & Herman, L. M. (1992). Humpback whale song: Our current understanding. In J. A. Thomas, R. A. Kastelein, & A. S. Supin (Eds.), *Marine mammal sensory systems* (pp. 459–483). Plenum.

Herman, L. M. (2010). What laboratory research has told us about dolphin cognition. *International Journal of Comparative Psychology, 23*, 310–330.

Kowarski, K., Cerchio, S., Whitehead, H., Cholewiak, D., & Moors-Murphy, H. (2023). Seasonal song ontogeny in western North Atlantic humpback whales: Drawing parallels with songbirds. *Bioacoustics, 32*, 325–347.

Mercado, E., III, Frazer, L. N., & Herman, L. M. (1996). Humpback whale sonar. *Journal of the Acoustical Society of America, 100*, 2644.

Mobley, J. R., Herman, L. M., & Frankel, A. S. (1988). Responses of wintering humpback whales (*Megaptera novaeangliae*) to playback of recordings of winter and summer vocalizations and of synthetic sound. *Behavioral Ecology and Sociobiology, 23*, 211–223.

Pack, A. A., Herman, E. Y., Baker, S. C., Bauer, G. B., Clapham, P. J., Connor, R. C., Craig, A. S., Forestell, P. H., Frankel, A. S., Notarbartolo di Sciara, G., Hoffmann-Kuhnt, M., Mercado, E., III, Mobley, J., Shyan-Norwalt, M. R., Spitz, S. S., Solangi, M., Thompson, R. K. R., von Fersen, L., Uyeyama, R., . . . Wolz, J. P. (2016). Louis M. Herman 1930-2016. *Marine Mammal Science, 33*, 389–406.

Tyack, P., & Whitehead, H. (1982). Male competition in large groups of wintering humpback whales. *Behaviour, 83*, 132–154.

Chapter 2. When Whales Sing

Clapham, P. J. (2000). The humpback whale: Seasonal breeding and feeding in a baleen whale. In J. Mann, R. C. Connor, P. L. Tyack, & H. Whitehead (Eds.),

Cetacean societies: Field studies of dolphins and whales (pp. 173–196). University of Chicago Press.

Lilly, J. C. (1977). *The deep self: Profound relaxation and the tank isolation technique.* Simon and Schuster.

Winter Wonder Coast

Au, W. W. L., Mobley, J., Burgess, W. C., Lammers, M. O., & Nachtigall, P. E. (2000). Seasonal and diurnal trends of chorusing humpback whales wintering in waters off western Maui. *Marine Mammal Science, 16,* 530–544.

Clark, C. W., & Clapham, P. J. (2004). Acoustic monitoring on a humpback whale (*Megaptera novaeangliae*) feeding ground shows continual singing into late spring. *Proceedings of Biological Sciences, 271,* 1051–1057.

Geist, V. (1971). *Mountain sheep: A study in behavior and evolution.* University of Chicago Press.

Gong, Z., Jain, A. D., Tran, D., Yi, D. H., Wu, F., Zorn, A., Ratilal, P., & Makris, N. C. (2014). Ecosystem scale acoustic sensing reveals humpback whale behavior synchronous with herring spawning processes and re-evaluation finds no effect of sonar on humpback song occurrence in the Gulf of Maine in fall 2006. *PLoS One, 9,* e104733.

Griffin, D. R. (1942). *Listening in the dark: The acoustic orientation of bats and men.* Yale University Press.

Herman, L. M., Pack, A. A., Spitz, S. S., Herman, E. Y., Rose, K., Hakala, S., & Deakos, M. H. (2013). Humpback whale song: Who sings? *Behavioral Ecology and Sociobiology, 67,* 1653–1663.

Morete, M. E., Freitas, A., Engel, M. H., Pace, R. M., III, & Clapham, P. J. (2003). A novel behavior observed in humpback whales on wintering grounds at Abrolhos Bank (Brazil). *Marine Mammal Science, 19,* 694–707.

Nielsen, R. O. (1991). *Sonar signal processing.* Artech House.

Nishimura, C. E. (1994). *Monitoring whales and earthquakes by using SOSUS.* Naval Research Laboratory.

Payne, K., & Payne, R. S. (1985). Large scale changes over 19 years in songs of humpback whales in Bermuda. *Zeitschrift fur Tierpsychologie, 68,* 89–114.

Rizzo, L. Y., & Schulte, D. (2009). A review of humpback whales' migration patterns worldwide and their consequences to gene flow. *Journal of the Marine Biological Association of the United Kingdom, 89,* 995–1002.

Schreiber, O. W. (1952). Some sounds from marine life in the Hawaiian area. *Journal of the Acoustical Society of America, 24,* 116.

Watkins, W. A., George, J. E., Daher, M. A., Mullin, K., Martin, D. L., Haga, S. H., & DiMarzio, N. A. (2000). *Whale call data for the north Pacific, November 1995 through July 1999, occurrence of calling whales and source locations from SOSUS and other acoustic systems.* Woodshole Oceanographic Institute, WHOI-00-02.

The Mind of the Humpback Whale

Branstetter, B. K., Finneran, J. J., Fletcher, E. A., Weisman, B. C., & Ridgway, S. H. (2012). Dolphins can maintain vigilant behavior through echolocation for 15 days without interruption or cognitive impairment. *PLoS One, 7,* e47478.

Fichtelius, K.-E., & Sjölander, S. (1972). *Smarter than man? Intelligence in whales, dolphins, and humans.* Pantheon.

Langworthy, O. R. (1932). A description of the central nervous system of the porpoise (*Tursiops truncatus*). *Journal of Comparative Neurology, 54,* 437–499.

Lilly, J. C. (1961). *Man and dolphin.* Doubleday & Company.

Mann, J. (Ed.). (2017). *Deep thinkers: Inside the minds of whales, dolphins, and porpoises.* University of Chicago Press.

Rattenborg, N. C., Amlaner, C. J., & Lima, S. L. (2000). Behavioral, neurophysiological and evolutionary perspectives on unihemispheric sleep. *Neuroscience and Biobehavioral Reviews, 24,* 817–842.

Chapter 3. Which Whales Sing

Girola, E., Dunlop, R. A., & Noad, M. J. (2023). Singing in a noisy ocean: Vocal plasticity in male humpback whales. *Bioacoustics, 32,* 301–324.

Narganes Homfeldt, T., Risch, D., Stevenson, A., & Henry, L. A. (2022). Seasonal and diel patterns in singing activity of humpback whales migrating through Bermuda. *Frontiers in Marine Science, 9,* 1764.

Big Boys with Big Brains

Atkinson, S., Branch, T. A., Pack, A. A., Straley, J. M., Moran, J. R., Gabriele, C., Mashburn, K. L., Cates, K., & Yin, S. (2023). Pregnancy rate and reproductive hormones in humpback whale blubber: Dominant form of progesterone differs during pregnancy. *General and Comparative Endocrinology, 330,* 114151.

Cabrera, A. A., Bérubé, M., Lopes, X. M., Louis, M., Oosting, T., Rey-Iglesia, A., Rivera-León, V. E., Székely, D., Lorenzen, E. D., & Palsbøll, P. J. (2021). A genetic perspective on cetacean evolution. *Annual Review of Ecology, Evolution, and Systematics, 52,* 131–151.

Cates, K. A., Atkinson, S., Gabriele, C. M., Pack, A. A., Straley, J. M., & Yin, S. (2019). Testosterone trends within and across seasons in male humpback whales (*Megaptera novaeangliae*) from Hawaii and Alaska. *General and Comparative Endocrinology, 279,* 164–173.

Clark, C. W., & Garland, E. C. (Eds.). (2022). *Ethology and behavioral ecology of mysticetes.* Springer.

Gatesy, J., Geisler, J. H., Chang, J., Buell, C., Berta, A., Meredith, R. W., Springer, M. S., & McGowen, M. R. (2013). A phylogenetic blueprint for a modern whale. *Molecular Phylogenetics and Evolution, 66,* 479–506.

Kolarik, A. J., Cirstea, S., Pardhan, S., & Moore, B. C. (2014). A summary of research investigating echolocation abilities of blind and sighted humans. *Hearing Research, 310,* 60–68.

McDonald, M. A., Mesnick, S. L., & Hildebrand, J. A. (2006). Biogeographic characterization of blue whale song worldwide: Using song to identify populations. *Journal of Cetacean Research and Management, 8,* 55–65.

McGowen, M. R., Tsagkogeorga, G., Álvarez-Carretero, S., Dos Reis, M., Struebig, M., Deaville, R., Jepson, P. D., Jarman, S., Polanowski, A., Morin, P. A., & Rossiter, S. J. (2020). Phylogenomic resolution of the cetacean tree of life using target sequence capture. *Systematic Biology, 69,* 479–501.

Racicot, R. (2022). Evolution of whale sensory ecology: Frontiers in nondestructive anatomical investigations. *Anatomical Record, 305,* 736–752.

Riebel, K., Odom, K. J., Langmore, N. E., & Hall, M. L. (2019). New insights from female bird song: Towards an integrated approach to studying male and female communication roles. *Biology Letters, 15,* 20190059.

Spector, D. A. (1994). Definition in biology: The case of "bird song." *Journal of Theoretical Biology, 168,* 373–381.

Swartz, S. L., Lang, A., Burdin, A., Calambokidis, J., Frouin-Mouy, H., Martínez-Aguilar, S., Rodríguez-González, F., Tenorio-Hallé, L., Thode, A., Urbán-Ramírez, J., & Weller, D. W. (2023). Gray whale sex, reproductive behavior, and social strategies. In B. Würsig & D. N. Orbach (Eds.), *Sex in cetaceans: Morphology, behavior, and the evolution of sexual strategies* (pp. 499–520). Springer.

Singing Toothed Whales and Bats?

Bain, D. E. (1986). Acoustic behavior of *Orcinus*: Sequences, periodicity, behavioral correlates and an automated technique for call classification. In B. C. Kirkevold & J. S. Lockard (Eds.), *Behavioural biology of killer whales* (pp. 335–371). Alan R. Liss.

Finneran, J. J. (2013). Dolphin "packet" use during long-range echolocation tasks. *Journal of the Acoustical Society of America, 133,* 1796–1810.

Mayberry, H. W., Faure, P. A., & Ratcliffe, J. M. (2019). Sonar strobe groups and buzzes are produced before powered flight is achieved in the juvenile big brown bat, *Eptesicus fuscus. Journal of Experimental Biology, 222,* jeb209163.

Mora, E. C., Ibáñez, C., Macías, S., Juste, J., López, I., & Torres, L. (2011). Plasticity in the echolocation inventory of *Mormopterus minutus* (Chiroptera, Molossidae). *Acta Chiropterologica, 13*, 179–187.

Parks, S. E., & Clark, C. W. (2008). Acoustic communication in mysticetes. *Bioacoustics, 17*, 45–47.

Pavan, G., Hayward, T. J., Borsani, J. F., Priano, M., Manghi, M., Fossati, C., & Gordon, J. (2000). Time patterns of sperm whale codas recorded in the Mediterranean Sea 1985-1996. *Journal of the Acoustical Society of America, 107*, 3487–3495.

Selbmann, A., Miller, P. J. O., Wensveen, P. J., Svavarsson, J., & Samarra, F. I. P. (2023). Call combination patterns in Icelandic killer whales (*Orcinus orca*). *Scientific Reports, 13*, 21771.

Smotherman, M., Knornschild, M., Smarsh, G., & Bohn, K. (2016). The origins and diversity of bat songs. *Journal of Comparative Physiology A: Neuroethology, Sensory, Neural, and Behavioral Physiology, 202*, 535–554.

Chapter 4. Where Whales Sing

Chapman, D. M. (2004). You can't get there from here: Shallow water sound propagation and whale localization. *Canadian Acoustics, 32*, 167–171.

Meyer, J. (2021). Environmental and linguistic typology of whistled languages. *Annual Review of Linguistics, 7*, 493–510.

Urick, R. J. (1996). *Principles of underwater sound.* 3rd ed. Peninsula.

Oceanic Thin Disks

Clark, C. W., & Gagnon, G. J. (2022). Baleen whale acoustic ethology. In C. W. Clark & E. C. Garland (Eds.), *Ethology and behavioral ecology of mysticetes* (pp. 11–44). Springer.

Ellison, W. T., Clark, C. W., & Bishop, G. C. (1987). Potential use of surface reverberation by bowhead whales, *Balaena mysticetus*, in under-ice navigation. *Report of the International Whaling Commission, 37*, 329–332.

Forrest, T. G., Miller, G. L., & Zagar, J. R. (1993). Sound propagation in shallow water: Implications for acoustic communication by aquatic animals. *Bioacoustics, 4*, 259–270.

George, J. C., Clark, C., Carroll, G. M., & Ellison, W. T. (1989). Observations on the ice-breaking and ice navigation behavior of migrating bowhead whales (*Balaena mysticetus*) near Point Barrow, Alaska, spring 1985. *Arctic, 42*, 24–30.

Kuperman, W. A., & Lynch, J. F. (2004). Shallow-water acoustics. *Physics Today, 57*, 55–61.

Mercado, E., III, & Frazer, L. N. (1999). Environmental constraints on sound transmission by humpback whales. *Journal of the Acoustical Society of America, 106,* 3004–3016.

Payne, R. S., & Webb, D. (1971). Orientation by means of long-range acoustic signaling in baleen whales. *Annals of the New York Academy of Sciences, 188,* 110–142.

Stimpert, A. K., Peavey, L. E., Friedlaender, A. S., & Nowacek, D. P. (2012). Humpback whale song and foraging behavior on an Antarctic feeding ground. *PLoS One, 7*(12), e51214.

Wahlberg, M., & Larsen, O. N. (2017). Propagation of sound. In C. Brown & T. Riede (Eds.), *Comparative bioacoustics: An overview* (pp. 62–119). Bentham Science.

Ecological Origins of Song Complexity

Au, W. W., Frankel, A., Helweg, D. A., & Cato, D. H. (2001). Against the humpback whale sonar hypothesis. *IEEE Journal of Oceanic Engineering, 26,* 295–300.

Bass, A. H., & Clark, C. W. (2003). The physical acoustics of underwater sound communication. In A. M. Simmons, A. N. Popper, & R. R. Fay (Eds.), *Acoustic communication* (pp. 15–64). Springer.

Eveland, K., Stevenson, R., Domski, P., Vanderelst, D., & Kloepper, L. (2023). Spatial differences in soundscape for bats on the edge versus the center of a bat swarm. *Proceedings of Meetings on Acoustics, 51,* 010001.

Jensen, F. B., Kuperman, W. A., Porter, M. B., Schmidt, H., & Tolstoy, A. (2011). *Computational ocean acoustics.* Springer.

Kügler, A., Lammers, M. O., Zang, E. J., & Pack, A. A. (2021). Male humpback whale chorusing in Hawai'i and its relationship with whale abundance and density. *Frontiers in Marine Science, 8,* 735664.

Mossbridge, J. A., & Thomas, J. A. (1999). An "acoustic niche" for Antarctic killer whale and leopard seal sounds *Marine Mammal Science, 15,* 1351–1357.

Padgham, M. (2004). Reverberation and frequency attenuation in forests—implications for acoustic communication in animals. *Journal of the Acoustical Society of America, 115,* 402–410.

Richardson, W. J., Greene Jr., C. R., Malme, C. I., & Thomson, D. H. (2013). *Marine mammals and noise.* Academic Press.

Chapter 5. What Whales Sing

Jarrell, R. (1977). *A bat is born: From the bat poet.* Doubleday.

Jung, K., Molinari, J., & Kalko, E. K. (2014). Driving factors for the evolution of species-specific echolocation call design in new world free-tailed bats (Molossidae). *PLoS One, 9,* e85279.

Obrist, M. K. (1995). Flexible bat echolocation: The influence of individual habitat and conspecifics on sonar signal design. *Behavioral Ecology and Sociobiology, 36*, 207–219.

A Morphing Mix of Rhythmic Tweets and Moos

Cholewiak, D., Sousa-Lima, R., & Cerchio, S. (2013). Humpback whale song hierarchical structure: Historical context and discussion of current classification issues. *Marine Mammal Science, 29*, E312–E332.

Darling, J. D., Acebes, J. M. V., Frey, O., Jorge Urban, R., & Yamaguchi, M. (2019). Convergence and divergence of songs suggests ongoing, but annually variable, mixing of humpback whale populations throughout the North Pacific. *Scientific Reports, 9*, 7002.

Darling, J. D., & Sousa-Lima, R. S. (2005). Songs indicate interaction between humpback whale (*Megaptera novaeangliae*) populations in the western and eastern south Atlantic ocean. *Marine Mammal Science, 21*, 557–566.

Green, S. R., Mercado, E., III, Pack, A. A., & Herman, L. M. (2011). Recurring patterns in the songs of humpback whales (*Megaptera novaeangliae*). *Behavioural Processes, 86*, 284–294.

Handel, S., Todd, S. K., & Zoidis, A. M. (2012). Hierarchical and rhythmic organization in the songs of humpback whales (*Megaptera novaeangliae*). *Bioacoustics, 21*, 141–156.

Kello, C. T., Bella, S. D., Mede, B., & Balasubramaniam, R. (2017). Hierarchical temporal structure in music, speech and animal vocalizations: Jazz is like a conversation, humpbacks sing like hermit thrushes. *Journal of the Royal Society Interface, 14*, 20170231.

Kershenbaum, A., Blumstein, D. T., Roch, M. A., Akcay, C., Backus, G., Bee, M. A., Bohn, K., Cao, Y., Carter, G., Casar, C., Coen, M., DeRuiter, S. L., Doyle, L., Edelman, S., Ferrer-i-Cancho, R., Freeberg, T. M., Garland, E. C., Gustison, M., Harley, H. E., . . . Zamora-Gutierrez, V. (2016). Acoustic sequences in non-human animals: A tutorial review and prospectus. *Biology Reviews, 91*, 13–52.

Mercado, E., III. (2020). Song morphing by humpback whales: Cultural or epiphenomenal? *Frontiers in Psychology, 11*, 574403.

Mercado, E., III, Ashour, M., & McAllister, S. (2022). Cognitive control of song production by humpback whales. *Animal Cognition, 25*, 1133–1149.

Mercado, E., III, Herman, L. M., & Pack, A. A. (2003). Stereotypical sound patterns in humpback whale songs: Usage and function. *Aquatic Mammals, 29*, 37–52.

Mercado, E., III, & Perazio, C. E. (2021). Similarities in composition and transformations of songs by humpback whales (*Megaptera novaeangliae*) over time and space. *Journal of Comparative Psychology, 135*, 28–50.

Mercado, E., III, & Perazio, C. E. (2022). All units are equal in humpback whale songs, but some are more equal than others. *Animal Cognition, 25*, 149–177.

Schneider, J. N., & Mercado, E., III. (2019). Characterizing the rhythm and tempo of sound production by singing whales. *Bioacoustics, 28*, 239–256.

Echoic Consequences of Patterned Unit Production

Allen, J. A., Garland, E. C., Dunlop, R. A., & Noad, M. J. (2019). Network analysis reveals underlying syntactic features in a vocally learnt mammalian display, humpback whale song. *Proceedings of the Royal Society B: Biological Sciences, 286*, 20192014.

Corcoran, A. J., & Moss, C. F. (2017). Sensing in a noisy world: Lessons from auditory specialists, echolocating bats. *Journal of Experimental Biology, 220*, 4554–4566.

Kaufman, A. B., Green, S. R., Seitz, A. R., & Burgess, C. (2012). Using a self-organizing map (SOM) and the hyperspace analog to language (HAL) model to identify patterns of syntax and structure in the songs of humpback whales. *International Journal of Comparative Psychology, 25*, 237–275.

Simmons, J. A., & Stein, R. A. (1980). Acoustic imaging in bat sonar: Echolocation signals and the evolution of echolocation. *Journal of Comparative Physiology, 135*, 61–84.

Suzuki, R., Buck, J. R., & Tyack, P. L. (2006). Information entropy of humpback whale songs. *Journal of the Acoustical Society of America, 119*, 1849–1866.

Chapter 6. How Whales Sing

Au, W. W. L., Pack, A. A., Lammers, M. O., Herman, L. M., Deakos, M. H., & Andrews, K. (2006). Acoustic properties of humpback whale songs. *Journal of the Acoustical Society of America, 120*, 1103–1110.

Cazau, D., Adam, O., Laitman, J. T., & Reidenberg, J. S. (2013). Understanding the intentional acoustic behavior of humpback whales: A production-based approach. *Journal of the Acoustical Society of America, 134*, 2268–2273.

Mercado, E., III, Schneider, J. N., Pack, A. A., & Herman, L. M. (2010). Sound production by singing humpback whales. *Journal of the Acoustical Society of America, 127*, 2678–2691

Bidirectional Blowing

Brinklov, S., Fenton, M. B., & Ratcliffe, J. M. (2013). Echolocation in oilbirds and swiftlets. *Frontiers in Physiology, 4*, 123.

Cranford, T. W., Elsberry, W. R., Van Bonn, W. G., Jeffress, J. A., Chaplin, M. S., Blackwood, D. J., Carder, D. A., Kamolnick, T., Todd, M. A., & Ridgway, S. H. (2011). Observation and analysis of sonar signal generation in the bottlenose dolphin (*Tursiops truncatus*): Evidence for two sonar sources. *Journal of Experimental Marine Biology and Ecology, 407,* 81–96.

Deacon, T. W. (1997). *The symbolic species: The co-evolution of language and the brain.* W. W. Norton.

Elemans, C. P., Rasmussen, J. H., Herbst, C. T., During, D. N., Zollinger, S. A., Brumm, H., Srivastava, K., Svane, N., Ding, M., Larsen, O. N., Sober, S. J., & Svec, J. G. (2015). Universal mechanisms of sound production and control in birds and mammals. *Nature Communications, 6,* 8978.

Gandilhon, N., Adam, O., Cazau, D., Laitman, J. T., & Reidenberg, J. S. (2014). Two new theoretical roles of the laryngeal sac of humpback whales. *Marine Mammal Science, 31,* 774–781.

Kremers, D., Jaramillo, M. B., Boye, M., Lemasson, A., & Hausberger, M. (2011). Do dolphins rehearse show-stimuli when at rest? Delayed matching of auditory memory. *Frontiers in Psychology, 2,* 386.

Mercado, E., III, Mantell, J. T., & Pfordresher, P. Q. (2014). Imitating sounds: A cognitive approach to understanding vocal imitation. *Comparative Cognition & Behavior Reviews, 9,* 1–57.

Reidenberg, J. S., & Laitman, J. T. (2007). Discovery of the low frequency sound source in mysticeti (baleen whales): Anatomical establishment of a vocal fold homolog. *Anatomical Record, 290,* 745–759.

Reiss, D., & McCowan, B. (1993). Spontaneous vocal mimicry and production by bottlenose dolphins (*Tursiops truncatus*): Evidence for vocal learning. *Journal of Comparative Psychology, 107,* 301–312.

Schevill, W. E., Backus, R. H., & Hersey, J. B. (1962). Sound production by marine animals. In M. N. Hill (Ed.), *The sea.* John Wiley and Sons.

Stimpert, A. K., Wiley, D. N., Au, W. W., Johnson, M. P., & Arsenault, R. (2007). "Megapclicks": Acoustic click trains and buzzes produced during night-time foraging of humpback whales (*Megaptera novaeangliae*). *Biology Letters, 3,* 467–470.

When Complexity Is Not So Complex

Allen, J. A., Garland, E. C., Dunlop, R. A., & Noad, M. J. (2018). Cultural revolutions reduce complexity in the songs of humpback whales. *Proceedings of the Royal Society B: Biological Sciences, 285,* 20182088.

Bruni, L. E., & Giorgi, F. (2015). Towards a heterarchical approach to biology and cognition. *Progress in Biophysics and Molecular Biology, 119,* 481–492.

Cazau, D., Adam, O., Aubin, T., Laitman, J. T., & Reidenberg, J. S. (2016). A study of vocal nonlinearities in humpback whale songs: From production mechanisms to acoustic analysis. *Scientific Reports, 6*, 31660.

Fletcher, E. A. (2007). Animal bioacoustics. In T. D. Rossing (Ed.), *Handbook of acoustics* (pp. 785–804). Springer.

Mercado, E., III, & Handel, S. (2012). Understanding the structure of humpback whale songs (L). *Journal of the Acoustical Society of America, 132*, 2947–2950.

Murray, A., Dunlop, R. A., Noad, M. J., & Goldizen, A. W. (2018). Stereotypic and complex phrase types provide structural evidence for a multi-message display in humpback whales. *Journal of the Acoustical Society of America, 143*, 980–994.

Parsons, E. C. M., Wright, A. J., & Gore, M. A. (2008). The nature of humpback whale (*Megaptera novaeangliae*) song. *Journal of Marine Animals and Their Ecology, 1*, 22–31.

Shinbrot, T., Grebogi, C., Wisdom, J., & Yorke, J. A. (1992). Chaos in a double pendulum. *American Journal of Physics, 60*, 491–499.

Chapter 7. Who Hears What

Darling, J. D., Meagan, E., & Nicklin, C. P. (2006). Humpback whale songs: Do they organize males during the breeding season? *Behaviour, 143*, 1051–1101.

Mercado, E., III. (2018). The sonar model for humpback whale song revised. *Frontiers in Psychology, 9*, 1156.

Thaler, L., De Vos, H., Kish, D., Antoniou, M., Baker, C. J., & Hornikx, M. C. J. (2019). Human click-based echolocation of distance: Superfine acuity and dynamic clicking behaviour. *Journal of the Association for Research in Otolaryngology, 20*, 499–510.

Tyack, P. L., & Clark, C. W. (2000). Communication and acoustic behavior of dolphins and whales. In W. W. L. Au, A. N. Popper, & R. R. Fay (Eds.), *Hearing by whales and dolphins* (pp. 156–224). Springer.

Sonic Self-Gratification

Bates, M. E., Stamper, S. A., & Simmons, J. A. (2008). Jamming avoidance response of big brown bats in target detection. *Journal of Experimental Biology, 211*, 106–113.

Blauert, J. (1997). *Spatial hearing: The psychophysics of human sound localization.* MIT Press.

Branstetter, B. K., & Mercado, E., III. (2006). Sound localization by cetaceans. *International Journal of Comparative Psychology, 19*, 26–61.

Cranford, T. W., & Krysl, P. (2015). Fin whale sound reception mechanisms: Skull vibration enables low-frequency hearing. *PLoS One, 10,* e0116222.

De Vreese, S., Orekhova, K., Morell, M., Gerussi, T., & Graic, J. M. (2023). Neuroanatomy of the cetacean sensory systems. *Animals, 14,* 66.

Ketten, D. R. (2000). Cetacean ears. In W. W. L. Au, R. R. Fay, & A. N. Popper (Ed.), *Hearing by whales and dolphins* (pp. 43–108). Springer.

Madsen, P. T., de Soto, N. A., Arranz, P., & Johnson, M. (2013). Echolocation in Blainville's beaked whales (*Mesoplodon densirostris*). *Journal of Comparative Physiology A, 199,* 451–469.

Mercado, E., III. (2016). Acoustic signaling by singing humpback whales (*Megaptera novaeangliae*): What role does reverberation play? *PLoS One, 11,* e0167277.

Mercado, E., III, Wisniewski, M. G., McIntosh, B., Guillette, L., Hahn, A. H., & Sturdy, C. (2017). Chickadee songs provide hidden clues to singers' locations. *Animal Behavior and Cognition, 4,* 301–313.

Mooney, T. A., Kaplan, M. B., & Lammers, M. O. (2016). Singing whales generate high levels of particle motion: Implications for acoustic communication and hearing? *Biology Letters, 12,* 20160381.

National Marine Mammal Foundation. (2024, March 20). *First successful hearing tests conducted with minke whales will improve conservation and protection of baleen whales globally.* https://www.nmmf.org/our-work/biologic -bioacoustic-research/minke-whale-hearing/.

Rehorek, S. J., Stimmelmayr, R., George, J. C., Suydam, R., McBurney, D. M., & Thewissen, J. G. M. (2019). Structure of the external auditory meatus of the bowhead whale (*Balaena mysticetus*) and its relation to their seasonal migration. *Journal of Anatomy, 234,* 201–215.

Ryabov, V. A. (2023). The role of asymmetry of the left and right external ear of bottlenose dolphin (*Tursiops truncatus*) in the spatial localization of sound. *Acoustical Physics, 69,* 119–131.

Wisniewski, M. G., Mercado, E., III, Gramann, K., & Makeig, S. (2012). Familiarity with speech affects cortical processing of auditory distance cues and increases acuity. *PLoS One, 7,* e41025.

Yamato, M., Ketten, D. R., Arruda, J., Cramer, S., & Moore, K. (2012). The auditory anatomy of the minke whale (*Balaenoptera acutorostrata*): A potential fatty sound reception pathway in a baleen whale. *Anatomical Record, 295,* 991–998.

A "Whale's Ear View" of Whale Songs

Beamish, P. (1978). Evidence that a captive humpback whale (*Megaptera novaeangliae*) does not use sonar. *Deep Sea Research, 25,* 469–472.

Dent, M. L., & Bee, M. A. (2018). Principles of auditory object formation by nonhuman animals. In H. Slabbekoorn, R. J. Dooling, A. N. Popper, & R. R. Fay (Eds.), *Effects of anthropogenic noise on animals* (pp. 47–82). Springer.

Ketten, D. R. (1997). Structure and function in whale ears. *Bioacoustics, 8*, 103–135.

Neuweiler, G., Metzner, W., Heilmann, U., Rübsamen, R., Eckrich, M., & Costa, H. H. (1987). Foraging behaviour and echolocation in the rufous horseshoe bat (*Rhinolophus rouxi*) of Sri Lanka. *Behavioral Ecology and Sociobiology, 20*, 53–67.

Schnitzler, H. U., & Denzinger, A. (2011). Auditory fovea and Doppler shift compensation: Adaptations for flutter detection in echolocating bats using CF-FM signals. *Journal of Comparative Physiology A, 197*, 541–559.

Simmons, J. A., Houser, D., & Kloepper, L. (2014). Localization and classification of targets by echolocating bats and dolphins. In *Biosonar* (pp. 169–193). Springer.

Chapter 8. For Whom The Whales Toll

Bateson, P., & Laland, K. N. (2013). Tinbergen's four questions: An appreciation and an update. *Trends in Ecology & Evolution, 28*, 712–718.

Mercado, E., III. (2022). Intra-individual variation in the songs of humpback whales suggests they are sonically searching for conspecifics. *Learning and Behavior, 50*, 456–481.

Miller, P. J., Johnson, M. P., & Tyack, P. L. (2004). Sperm whale behaviour indicates the use of echolocation click buzzes "creaks" in prey capture. *Proceedings of the Royal Society of London. Series B: Biological Sciences, 271*, 2239–2247.

Surlykke, A., Nachtigall, P. E., Fay, R. R., & Popper, A. N. (Eds.). (2014). *Biosonar.* Springer.

Watkins, W. A. (1980). Acoustics and the behavior of sperm whales. In R.-G. Busnel & J. F. Fish (Eds.), *Animal sonar systems* (pp. 283–290). Springer.

Using Songs to Coordinate Connections

Bradbury, J. W. (1977). Lek mating behavior in the hammer-headed bat. *Zeitschrift für Tierpsychologie, 45*, 225–255.

Clapham, P. J. (1996). The social and reproductive biology of humpback whales: An ecological perspective. *Mammal Review, 26*, 27–49.

Clark, C. W., & Garland, E. C. (Eds.). (2022). *Ethology and behavioral ecology of mysticetes.* Springer.

Craig, A. S., Herman, L. M., Pack, A. A., & Waterman, J. O. (2014). Habitat segregation by female humpback whales in Hawaiian waters: Avoidance of males? *Behaviour, 151*, 613–631.

Darling, J. D., & Berube, M. (2001). Interactions of singing humpback whales with other males. *Marine Mammal Science, 17*, 570-584.

Darling, J. D., Jones, M. E., & Nicklin, C. P. (2012). Humpback whale (*Megaptera novaeangliae*) singers in Hawaii are attracted to playback of similar song (L). *Journal of the Acoustical Society of America, 132*, 2955-2958.

Frankel, A. S., Clark, C. W., Herman, L. M., & Gabriele, C. M. (1995). Spatial distribution, habitat utilization, and social interactions of humpback whales, *Megaptera novaeangliae*, off Hawai'i, determined using acoustic and visual techniques. *Canadian Journal of Zoology, 73*, 1134-1146.

Henderson, E. E., Helble, T. A., Ierley, G., & Martin, S. (2018). Identifying behavioral states and habitat use of acoustically tracked humpback whales in Hawaii. *Marine Mammal Science, 34*, 701-717.

Herman, L. M., & Tavolga, W. N. (1980). The communication systems of cetaceans. In L. M. Herman (Ed.), *Cetacean behavior: Mechanisms and functions* (pp. 149-209). Wiley Interscience.

Hogg, J. T. (1987). Intrasexual competition and mate choice in Rocky Mountain bighorn sheep. *Ethology, 75*, 119-144.

Kugler, A., Lammers, M. O., Pack, A. A., Tenorio-Halle, L., & Thode, A. M. (2024). Diel spatio-temporal patterns of humpback whale singing on a high-density breeding ground. *Royal Society Open Science, 11*, 230279.

Mastick, N. C., Wiley, D., Cade, D. E., Ware, C., Parks, S. E., & Friedlaender, A. S. (2022). The effect of group size on individual behavior of bubble-net feeding humpback whales in the southern Gulf of Maine. *Marine Mammal Science, 38*, 959-974.

Tyack, P., & Whitehead, H. (1982). Male competition in large groups of wintering humpback whales. *Behaviour, 83*, 132-154.

Social Learning and Vocal Culture in Singers

Cerchio, S., Jacobsen, J. K., & Norris, T. F. (2001). Temporal and geographical variations in songs of humpback whales, *Megaptera novaeangliae*: Synchronous change in Hawaiian and Mexican breeding assemblages. *Animal Behaviour, 62*, 313-329.

Fitch, W. T., & Jarvis, E. D. (2013). Birdsong and other animal models for human speech, song, and vocal learning. In M. A. Arbib (Ed.), *Language, music, and the brain: A mysterious relationship* (Vol. 10, pp. 499-539). MIT Press.

Garland, E. C., Goldizen, A. W., Rekdahl, M. L., Constantine, R., Garrigue, C., Hauser, N. D., Poole, M. M., Robbins, J., & Noad, M. J. (2011). Dynamic horizontal cultural transmission of humpback whale song at the ocean basin scale. *Current Biology, 21*, 687-691.

Garland, E. C., & McGregor, P. K. (2020). Cultural transmission, evolution, and revolution in vocal displays: Insights from bird and whale song. *Frontiers in Psychology, 11,* 544929.

Guinee, L. N., Chu, K., & Dorsey, E. M. (1983). Changes over time in the songs of known individual humpback whales (*Megaptera novaeangliae*). In R. S. Payne (Ed.), *Communication and behavior of whales* (pp. 59–80). Westview Press.

Mercado, E., III. (2022). The humpback's new songs: Diverse and convergent evidence against vocal culture via copying in humpback whales. *Animal Behavior and Cognition, 9,* 196–206.

Mercado, E., III, Herman, L. M., & Pack, A. A. (2005). Song copying by humpback whales: Themes and variations. *Animal Cognition, 8,* 93–102.

Panova, E. M., & Agafonov, A. V. (2017). A beluga whale socialized with bottlenose dolphins imitates their whistles. *Animal Cognition, 20,* 1153–1160.

Payne, K. (2000). The progressively changing songs of humpback whales: A window on the creative process in a wild animal. In N. L. Wallin, B. Merker, & S. Brown (Eds.), *Origins of music* (pp. 135–150). MIT Press.

Payne, K., Tyack, P., & Payne, R. S. (1983). Progressive changes in the songs of humpback whales (*Megaptera novaeangliae*): A detailed analysis of two seasons in Hawaii. In R. Payne (Ed.), *Communication and behavior of whales* (pp. 9–57). Westview Press.

Payne, R. B. (1985). Behavioral continuity and change in local song populations of village indigobirds *Vidua chalybeate*. *Zeitschrift für Tierpsychologie, 70,* 1–44.

Tchernichovski, O., & Marcus, G. (2014). Vocal learning beyond imitation: Mechanisms of adaptive vocal development in songbirds and human infants. *Current Opinion in Neurobiology, 28,* 42–47.

West, M. J., King, A. P., & White, D. J. (2003). Discovering culture in birds: The role of learning and development. In F. B. M. de Waal & P. L. Tyack (Eds.), *Animal social complexity: Intelligence, culture, and individualized societies* (pp. 470–494). Harvard University Press.

Chapter 9. Within Whales' Heads

Glezer, I. I., Jacobs, M. S., & Morgane, P. J. (1988). Implications of the "initial brain" concept for brain evolution in Cetacea. *Behavioral and Brain Sciences, 11,* 75–89.

Manger, P. R. (2006). An examination of cetacean brain structure with a novel hypothesis correlating thermogenesis to the evolution of a big brain. *Biological Reviews, 81,* 293–338.

Pitman, R. L., Deecke, V. B., Gabriele, C. M., Srinivasan, M., Black, N., Denkinger, J., Durban, J. W., Mathews, E. A., Matkin, D. R., Neilson, J. L., & Schulman-Janiger, A. (2017). Humpback whales interfering when mammal-eating killer whales attack other species: Mobbing behavior and interspecific altruism? *Marine Mammal Science, 33*, 7-58.

Schnupp, J., Nelken, I., & King, A. P. (2011). *Auditory neuroscience: Making sense of sound.* MIT Press.

Belittled Brains

Breathnach, A. S. (1955). The surface features of the brain of the humpback whale (*Megaptera novaeangliae*). *Journal of Anatomy, 89*, 343-354.

Butti, C., Janeway, C. M., Townshend, C., Wicinski, B. A., Reidenberg, J. S., Ridgway, S. H., Sherwood, C. C., Hof, P. R., & Jacobs, B. (2015). The neocortex of cetartiodactyls: I. A comparative Golgi analysis of neuronal morphology in the bottlenose dolphin (*Tursiops truncatus*), the minke whale (*Balaenoptera acutorostrata*), and the humpback whale (*Megaptera novaeangliae*). *Brain Structure and Function, 220*, 3339-3368.

Herculano-Houzel, S., Avelino-de-Souza, K., Neves, K., Porfirio, J., Messeder, D., Mattos Feijo, L., Maldonado, J., & Manger, P. R. (2014). The elephant brain in numbers. *Frontiers in Neuroanatomy, 8*, 46.

Hof, P. R., Chanis, R., & Marino, L. (2005). Cortical complexity in cetacean brains. *Anatomical Record A: Discoveries in Molecular, Cellular, and Evolutionary Biology, 287*, 1142-1152.

Hof, P. R., & Van der Gucht, E. (2007). Structure of the cerebral cortex of the humpback whale, *Megaptera novaeangliae* (Cetacea, Mysticeti, Balaenopteridae). *Anatomical Record, 290*, 1-31.

Horn, A. G., Leonard, M. L., Ratcliffe, L. M., Shackleton, S. A., & Weisman, R. (1992). Frequency variation in the songs of black-capped chickadees (*Parus atricapillus*). *Auk, 109*, 847-852.

Janik, V. M. (2014). Cetacean vocal learning and communication. *Current Opinion in Neurobiology, 28*, 60-65.

King, S. L., Guarino, E., Keaton, L., Erb, L., & Jaakkola, K. (2016). Maternal signature whistle use aids mother-calf reunions in a bottlenose dolphin, *Tursiops truncatus. Behavioural Processes, 126*, 64-70.

Ming, C., Haro, S., Simmons, A. M., & Simmons, J. A. (2021). A comprehensive computational model of animal biosonar signal processing. *PLoS Computational Biology, 17*, e1008677.

Morton, E. S. (1996). Why songbirds learn songs: An arms race over ranging? *Poultry and Avian Biology Reviews, 7*, 65-71.

Moss, C. F., & Sinha, S. R. (2003). Neurobiology of echolocation in bats. *Current Opinion in Neurobiology, 13*, 751–758.

Pollak, G. D., & Casseday, J. H. (2012). *The neural basis of echolocation in bats.* Springer Science & Business Media.

Raghanti, M. A., Wicinski, B., Meierovich, R., Warda, T., Dickstein, D. L., Reidenberg, J. S., Tang, C. Y., George, J. C., Hans Thewissen, J. G. M., Butti, C., & Hof, P. R. (2019). A comparison of the cortical structure of the bowhead whale (*Balaena mysticetus*), a basal mysticete, with other cetaceans. *Anatomical Record, 302*, 745–760.

Rice, A., Sirovic, A., Hildebrand, J. A., Wood, M., Carbaugh-Rutland, A., & Baumann-Pickering, S. (2022). Update on frequency decline of Northeast Pacific blue whale (*Balaenoptera musculus*) calls. *PLoS One, 17*, e0266469.

Siletti, K., Hodge, R., Mossi Albiach, A., Lee, K. W., Ding, S. L., Hu, L., Lonnerberg, P., Bakken, T., Casper, T., Clark, M., Dee, N., Gloe, J., Hirschstein, D., Shapovalova, N. V., Keene, C. D., Nyhus, J., Tung, H., Yanny, A. M., Arenas, E., . . . Linnarsson, S. (2023). Transcriptomic diversity of cell types across the adult human brain. *Science, 382*, eadd7046.

Smolker, R. A., Mann, J., & Smuts, B. B. (1993). Use of signature whistles during separations and reunions by wild bottlenose dolphin mothers and infants. *Behavioral Ecology and Sociobiology, 33*, 393–402.

Vernes, S. C., & Wilkinson, G. S. (2020). Behaviour, biology and evolution of vocal learning in bats. *Philosophical Transactions of the Royal Society B, 375*, 20190061.

Wund, M. A. (2005). Learning and the development of habitat-specific bat echolocation. *Animal Behaviour, 70*, 441–450.

The Ever-Changing Cells of Whales

Dell, L. A., Karlsson, K. A., Patzke, N., Spocter, M. A., Siegel, J. M., & Manger, P. R. (2016). Organization of the sleep-related neural systems in the brain of the minke whale (*Balaenoptera acutorostrata*). *Journal of Comparative Neurology, 524*, 2018–2035.

Heffner, R. S., Koay, G., Heffner, H. E., & Mason, M. J. (2022). Hearing in African pygmy hedgehogs (*Atelerix albiventris*): Audiogram, sound localization, and ear anatomy. *Journal of Comparative Physiology A, 208*, 653–670.

Irvine, D. R. F. (2018). Plasticity in the auditory system. *Hearing Research, 362*, 61–73.

Kalko, E. K., & Schnitzler, H. U. (1993). Plasticity in echolocation signals of European pipistrelle bats in search flight: Implications for habitat use and prey detection. *Behavioral Ecology and Sociobiology, 33*, 415–428.

Mercado, E., III. (2021). Spectral interleaving by singing humpback whales: Signs of sonar. *Journal of the Acoustical Society of America, 149,* 800–806.

Smith, A. B., Madsen, P. T., Johnson, M., Tyack, P., & Wahlberg, M. (2021). Toothed whale auditory brainstem responses measured with a non-invasive, on-animal tag. *JASA Express Letters, 1,* 091201.

Suga, N. (2020). Plasticity of the adult auditory system based on corticocortical and corticofugal modulations. *Neuroscience and Biobehavioral Reviews, 113,* 461–478.

Chapter 10. Will Whales Sing?

Bernasconi, M., Patel, R., Nottestad, L., Pedersen, G., & Brierley, A. S. (2013). The effect of depth on the target strength of a humpback whale (*Megaptera novaeangliae*). *Journal of the Acoustical Society of America, 134,* 4316.

Human-Whale-Science Relationships: It's Complicated

Blair, H. B., Merchant, N. D., Friedlaender, A. S., Wiley, D. N., & Parks, S. E. (2016). Evidence for ship noise impacts on humpback whale foraging behaviour. *Biology Letters, 12,* 20160005.

Erbe, C., Marley, S. A., Schoeman, R. P., Smith, J. N., Trigg, L. E., & Embling, C. B. (2019). The effects of ship noise on marine mammals—A review. *Frontiers in Marine Science, 6,* 606.

Fournet, M. E., Matthews, L. P., Gabriele, C. M., Haver, S., Mellinger, D. K., & Klinck, H. (2018). Humpback whales *Megaptera novaeangliae* alter calling behavior in response to natural sounds and vessel noise. *Marine Ecology Progress Series, 607,* 251–268.

Hazen, E. L., Abrahms, B., Brodie, S., Carroll, G., Jacox, M. G., Savoca, M. S., Scales, K. L., Sydeman, W. J., & Bograd, S. J. (2019). Marine top predators as climate and ecosystem sentinels. *Frontiers in Ecology and the Environment, 17,* 565–574.

Pacheco, A. S., Sepúlveda, M., & Corkeron, P. (2021). Whale-watching impacts: Science, human dimensions and management. *Frontiers in Marine Science, 8,* 737352.

Sellheim, N. (2020). *International marine mammal law.* Springer.

Sprogis, K. R., Videsen, S., & Madsen, P. T. (2020). Vessel noise levels drive behavioural responses of humpback whales with implications for whale-watching. *Elife, 9,* e56760.

Tsujii, K., Akamatsu, T., Okamoto, R., Mori, K., Mitani, Y., & Umeda, N. (2018). Change in singing behavior of humpback whales caused by shipping noise. *PLoS One, 13,* e0204112.

What Would Be Lost?

Albouy, C., Delattre, V., Donati, G., Frolicher, T. L., Albouy-Boyer, S., Rufino, M., Pellissier, L., Mouillot, D., & Leprieur, F. (2020). Global vulnerability of marine mammals to global warming. *Scientific Reports, 10,* 548.

Cabrera, A. A., Schall, E., Berube, M., Anderwald, P., Bachmann, L., Berrow, S., Best, P. B., Clapham, P. J., Cunha, H. A., Dalla Rosa, L., Dias, C., Findlay, K. P., Haug, T., Heide-Jorgensen, M. P., Hoelzel, A. R., Kovacs, K. M., Landry, S., Larsen, F., Lopes, X. M., . . . Palsboll, P. J. (2022). Strong and lasting impacts of past global warming on baleen whales and their prey. *Global Change Biology, 28,* 2657–2677.

Godfrey-Smith, P. (2020). *Metazoa: Animal life and the birth of the mind.* Farrar, Straus and Giroux.

Kershaw, J. L., Ramp, C. A., Sears, R., Plourde, S., Brosset, P., Miller, P. J. O., & Hall, A. J. (2020). Declining reproductive success in the Gulf of St. Lawrence's humpback whales (*Megaptera novaeangliae*) reflects ecosystem shifts on their feeding grounds. *Global Change Biology, 27,* 1027–1041.

Papagianni, D., & Morse, M. A. (2015). *The Neanderthals rediscovered: How modern science is rewriting their story.* Thames & Hudson.

Tulloch, V. J. D., Plaganyi, E. E., Brown, C., Richardson, A. J., & Matear, R. (2019). Future recovery of baleen whales is imperiled by climate change. *Global Change Biology, 25,* 1263–1281.

van Weelden, C., Towers, J. R., & Bosker, T. (2021). Impacts of climate change on cetacean distribution, habitat and migration. *Climate Change Ecology, 1,* 100009.

von Hammerstein, H., Setter, R. O., van Aswegen, M., Currie, J. J., & Stack, S. H. (2022). High-resolution projections of global sea surface temperatures reveal critical warming in humpback whale breeding grounds. *Frontiers in Marine Science, 9,* 837772.

INDEX

dissertation, 23, 71–72, 83
dive cycle, 140–41, 152
dolphin cognition, 14–15, 71, 120–21
dolphins, vii, xi; air sacs, 131; calves, 228–30; dentition, 56; dreams, 44; eavesdropping, 189; effects of noise on, 120; language, 43, 197–98; minds, 120, 212; number concepts, 20; vocalizations, 9–18, 63, 94, 128, 136, 199–203; whale imitations, 133–35
dolphin song, 47, 61, 63–64, 79, 135, 267
donkeys, 132–33, 140, 231
Doppler shift, 164, 168
double pendulum, 140–42
double slit experiment, 90–91

eardrum, 152
ears, ix, 1, 26, 151–57, 162–70, 253–54, 275
eavesdropping, 12, 149, 174, 181, 229, 262
echoic scene, 110, 113, 119, 211, 246–47
echoic soundscape, 249, 274
echoic streams, 11, 25, 96, 102–3, 107, 110, 112–13, 119, 149, 164, 169–70, 174–75, 211, 224–25, 263
echolocation: bat, ix–x, 10–13, 29–30, 122, 175–79, 252, 273–81; beluga, 17–18, 63, 166–67, 197–98, 250; bird, ix, 57; dolphin, 10, 54–57, 64, 98, 121, 128, 136, 147–48, 162–65, 276–79; ecosystems, 210, 251, 258; human, ix, 57; humpback whale, ix, 268
elaborate traits, 48, 111
electrophysiology, 221–23, 243
encephalization quotient, 217; Endangered Species Act, 254
escort, 186–87, 190, 192
ethologists, 177
evolution: bat baby making, 48; environmental factors in, 85–87, 251–55, 273, 277–79; favoring song learning, 226, 230; favoring vocal stability, 94; forces shaping brain circuits, 221; hypothesized mechanisms driving song form, 18, 66, 127, 197–202; leading to click-based echolocation, 136; split between toothed and baleen whales, 55–57, 153; within reverberant habitats, 164, 240

ewes, 37, 52, 111, 190–91. *See also* sheep
experimentum crucis, 90–91

fee-bee song, 159–60, 162, 225–26, 232. *See also* chickadees
feeding ground, 31–36
feeding singers, 76–78
female singers, 49, 60–61
Fichtelius, Karl-Erik, 43
field studies, 202–3
fish, 11, 76, 113, 135, 157, 181–83, 213, 234
Frankel, Adam, 16
Frazer, Neil, 18–23, 71, 81–84
fused ear bones, 155

geophysics, 18
glove finger, 152
Griffin, Donald, 30, 252, 260–61, 273, 276, 280

Hauser, Nan, 212–13
Hawai'i, 14–16, 31, 71–72, 76, 108, 183, 196, 202–6; heards of whales, 180–81
hearing. *See* ears; hearing sensitivities
hearing sensitivities, 120, 151, 167
hedgehogs, 42–43, 214, 216, 244
Heidegger, viii–ix
helium, 105, 139
Herman, Lou, 15–16, 20–23, 71, 88, 120, 161, 184, 192, 202–4
hierarchical structure, 101–4
hippopotamuses, 36
Historia Animalium (book), vii
holy grail, 91
homing beacon, 88, 189, 229
horses, 141
horseshoe bat, 163–70, 173–74, 240
human echolocation, ix, 57
humpback whales, 11, 28, 32–35, 136, 190–93, 212–13
Humpback Whale World Congress, 161
hydrophone, 29–30, 76, 150

IBM, 14–16, 25
infrasonic, 59, 155
insect song, 4, 63
integration time window, 223–24

internal air recirculation, 140. *See also* bidirectional sound production

inter-unit interval, 94–95, 97, 103, 106–7, 110, 143, 224

jaws, 153, 155
jhanas, 263
Jung, Carl, 262

Kewalo Basin Marine Mammal Laboratory, 14–15, 20, 25, 120
kittens, 274
krill, 36

language learning, 15, 43, 198
Langworthy, Orthello, 41–42, 216
laryngeal sac, 131–32
learning to hear, ix, 211
lek, 184–85, 192
lekking hypothesis, 184–85
Lilly, John, 27, 43–44, 197, 215
little goblin bat, 64–65, 83, 95, 240
Loewi, Otto, 17
lung power, 126

Madagascar, 160–62
Man and Dolphin (book), 43
marine mammalogists, 147
Marine Mammal Protection Act, 254–55
mating arena, 184. *See also* lek
mating display, 7, 48
McVay, Scott, 100, 267
megalohydrothalassaphobia, 262
megapclicks, 122, 125
Megaptera novaeangliae, xi, 54. *See also* humpback whales
mental states, viii, 43, 147, 212, 217, 260–61
meows, 34, 225
migration, 33, 76–79, 258
milling, 183
mind, 27–28, 40, 120, 194, 213, 259–64, 267–69
Moby Dick (book), 176
morphing, 98, 103–7, 110, 114–15, 142, 202–5, 210, 224, 231, 236, 239, 247–48
Morton, Eugene, 226
motionlessness, 126, 143

music, 24–26, 31, 58, 60, 62, 67, 83, 88, 110, 118, 167–69, 243–45
Mysticeti, x–xi

navigation, 29, 79–80, 266; Neanderthals, 260
neuroanatomy, 41–42, 165–67, 216–19, 237
neurons, 41–42, 90, 164–70, 174, 215–23, 227, 233–38, 243
Newton, Isaac, 17–18
nightingales, 82
noise, 85–86, 114, 120, 150–51, 160, 172, 245, 255–61
Norris, Ken, 252
nose lips, 129
number concepts, 20

ocean acoustics, 18–19, 70–77, 80–91, 173, 236, 253, 258
ocean donkeys, 132–33, 231
oceanic telephone game, 196
ocean temperature, 73, 258–59
orcas, 24, 63–64, 120, 157, 181, 213, 228, 235, 267
orca song, 63–64, 267

Paynes, Katy and Roger, 1–2, 5, 30–31, 93, 100–101, 267
peacocks, 5–7, 111, 127, 193, 249
phrases, 25, 101–6, 109–15, 134, 139, 142–43, 160–62, 169–74, 200, 209–11, 241, 246–48
playback experiments, 156, 159, 237–38
plip-plop calls, 96–100, 105–10, 142–43
polygynous, 54
Portuguese man-of-war, 250
prairie dog, 4
progressive changes in songs, 108, 205, 231
promiscuity, 54
Puerto Rico, 2, 205–6

Quasimodean humans, 260

RADAR, 29
rams, 37–38, 40, 52–53, 111, 190, 191. *See also* sheep
ranging, 29, 77–79, 274

ranging hypothesis, 226-30, 233, 238
ranting, 22, 265
raptors, 118, 179, 234-35
rats, 72, 215, 236, 243
recurring themes, 205
Reidenberg, Joy, 129, 131
reproductive display, 62-64, 194, 197
resonant air cavities, 139
reverberation, 85-86, 158-64, 169-70, 226, 232-34, 245, 253
rhythm, 2, 63-65, 82, 95-110, 134, 139-43, 206
Ridgway, Sam, 198

scientific journals, 23, 72, 81-83
screaming baby, 63
self-interference, 110, 242
sensitive wings, 275
sensory deprivation chamber, 27-28
sexually receptive females, 50-51, 65
sexual selection, 111
sheep, 26, 36-37, 51-53, 111, 190-91, 235
shifting units, 102-3, 211. See also morphing
shipping traffic, 254-55, 258
shore-based scan, 180
signature whistles, 199, 228-30
silbo, 69-70
singerings, 246-47
sirens, 2, 88-89, 171
Skellington, Jack, 208
skull, 153-55, 167, 215, 248, 279
Smarter than Man? (book), 43
social context, 136, 158-59, 198
social learning, 194
SONAR, 29
sonar hypothesis, 9-10, 14, 19-23, 46-48, 64, 75, 81-84, 94, 109, 115, 119, 124-25, 143, 146, 173, 177, 193, 221-23, 238-42, 248, 252-54, 265
Sonar of Dolphins, The (book), 17
songbirds, 8-10, 40, 59-65, 82, 137, 158-60, 194-202, 225-27, 234-35
song copying, 196-211, 204, 230-31, 249. See also vocal imitation
song distortion, 66, 87, 90-94
song-generated echoes, 162. See also sonar hypothesis

song learning, 59-61, 194-200, 226-33, 239. See also vocal learning
song novelty/innovations, 173, 205-10, 208, 225; song repertoires, 8, 200
song session, 104-6, 115, 136, 190, 241, 246-47
song spread, 196. See also cultural evolution
sound channel, 74, 87, 152-55, 173
sound localization, 89, 95, 156-72, 187, 227-30, 265, 275; sound pattern, 4, 34, 80, 97-101, 107-8, 135, 140-43, 146, 197-98. See also phrases
sound propagation, 18, 73-77, 80-94, 113, 129, 157-63, 173, 258; spacing hypothesis, 88, 185, 192
Spallanzani, Lazaro, 30
sperm whales, 43, 62-63, 80, 176-78
Swift, Taylor, 181
swiftlets, ix, 57
syntax, 118, 291

template, 149, 174
tending bond, 191-92
territories, 7, 32, 82, 88, 137, 158-59, 184-85, 201
testosterone, 50
thalassaphobia, 262
theme cycles, 101, 110, 115, 247-48
themes, 101-6, 110, 115, 141, 178, 204-6, 211, 247-48; Thousand Mile Song (book), 82
time delays, 222
Tinbergen, Niko, 177
toothed whales: brains of, 216, 253; clicks of, 210; dogmatic beliefs about, 65; singing in, 61-64, 267; sound production in, 121, 125, 128, 225, 279; sound reception in, 57, 151, 153, 155-56, 165-68; species of, 54-55; vocal evolution in, 251; vocal imitation in, 198-200, 230; transitional phrases, 106, 247
Tyack, Peter, 8, 187

U-folds, 131
ultrasound/ultrasonic, viii, x, 30, 55-57, 65, 90, 95, 142-47, 153-56, 223, 276-80; unihemispheric sleep, 44, 242-43

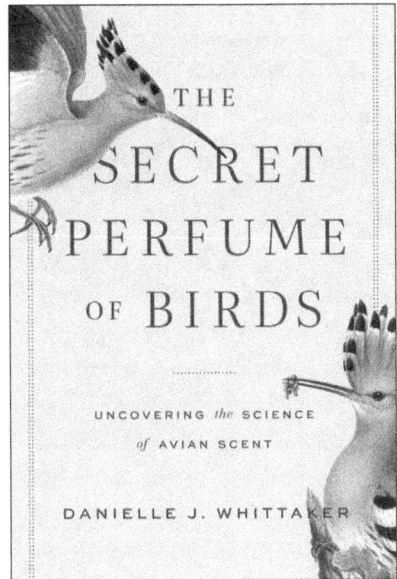

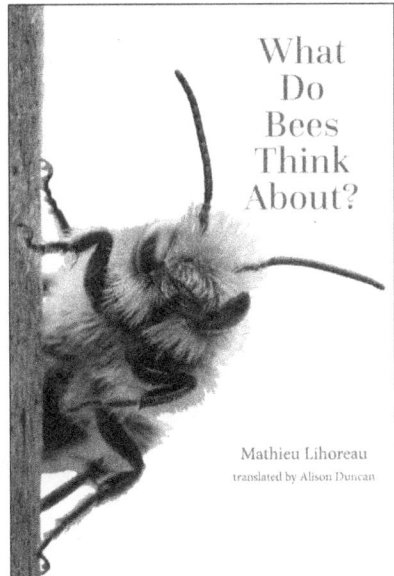